P9-ARG-895

HVAC Controls
and Control Systems

HVAC Controls and Control Systems

DON SWENSON

REGENTS/PRENTICE HALL, Englewood Cliffs, NJ 07632

Library of Congress Cataloging-in-Publication Data

Swenson, Don.
   HVAC Controls and control systems / Don Swenson.
     p.   cm.
   Includes index.
   ISBN (invalid) 0-13-045360-7
   1. Heating—Control.   2. Air conditioning—Control.   I. Title.
TH7466.5.S93 1994
697—dc20

Production Editor: *Marcia Krefetz*
Acquisitions Editor: *Ed Francis*
Prepress Buyer: *Ilene Sanford*
Manufacturing Buyer: *Ed O'Dougherty*
Editorial Assistant: *Gloria Schaffer*

1994 by REGENTS/PRENTICE HALL
A Division of Simon & Schuster
Englewood Cliffs, New Jersey 07632

Printed in the United States of America

10 9 8 7 6 5 4 3 2 1

ISBN: 0-13-045360-9

Prentice-Hall International (UK) Limited, *London*
Prentice-Hall of Australia Pty. Limited, *Sydney*
Prentice-Hall Canada Inc., *Toronto*
Prentice-Hall Hispanoamericana, S.A., *Mexico*
Prentice-Hall of India Private Limited, *New Delhi*
Prentice-Hall of Japan, Inc., *Tokyo*
Simon & Schuster Asia Pte. Ltd., *Singapore*
Editora Prentice-Hall do Brasil, Ltda., *Rio de Janeiro*

Contents

Preface

There are many good control books available today, but most of them are highly proprietary. The books usually cover one type of control system in detail but have little information about the other systems. Most control books divorce the controls from the equipment they control. This book is written from a highly practical viewpoint, with special emphasis on the application of the controls.

To be most useful a textbook should present information that can be applied to real-world situations. This book can be used by both students and technicians already in the field. It integrates the learning process with application of the information learned. It is presented in a systematic way in which the reader moves from simpler control systems to those that are more complex.

The book is organized to cover the reasons control systems are used, the general theory of control systems operation, electric controls systems and applications, pneumatic control systems and operations, electronic control systems and operations, and finally, the use of electronic controls combined with computerized systems to provide total building system control.

Don Swenson

1

Introduction to HVAC Control Systems

An *HVAC control system* is a combination of control devices and the network of control signal communication lines that tie the various units of an HVAC system together and control the units so that they produce a comfortable climate in a building. Controls and control systems have been used to operate heating, ventilating, and air-conditioning (HVAC) systems almost as long as the systems have existed. Heating living spaces to make them more comfortable is as old as people's first use of fire. Cavemen standing around a fire, in the open or in a cave, were attempting to control their environment by heating it. Of course, control of the heat was minimal and was entirely manual. Another log was thrown on the fire if more heat was needed. If less heat was needed, the fire was allowed to die down.

Ventilation of living spaces to make them more habitable has also been practiced since human beings moved into closed spaces. Outside air was then, and still is, the best agent to use for keeping air in closed spaces fresh. In older buildings, if the air in a building became stale, the windows or doors were opened to let outdoor air in. Today, we use blowers and sophisticated duct systems to bring the air in.

HEATING CONTROLS

When centralized heating units and systems were first used on a large scale in the late nineteenth century, heating units were controlled manually. Heating output of the units was controlled by controlling the amount of fuel fed to the furnace and by adjustment of the dampers on the furnace. As heating systems became more common, the need for automatic control of the systems became more obvious.

Many of the first large buildings that had central heating systems installed used steam or hot water as a source of heat. This is called a *hydronic system*. In this system steam or hot water circulated by gravity through a piping system and radiators in different rooms provided heat for those rooms. Control of the heating medium was by hand valves in the circulating pipes or on the radiators (Figure 1-1). This control was marginal, to say the least. If the valves were opened to provide the maximum amount of heat, the heated spaces were comfortable on cold days but overheated on mild days. If the valves were adjusted to provide enough heat when the weather was warmer, there was not enough heat when it was cold. Constant adjustment of the valves was not practical because of the time and labor required.

One of the first automatic control systems used for larger heating systems was a pneumatic system (Figure 1-2). Low-pressure air was used as a control medium. Thermostats located in the spaces to be controlled changed the control medium pressure in response to room temperatures. Piston or diaphragm operators positioned valves or dampers to provide the proper amount of heating

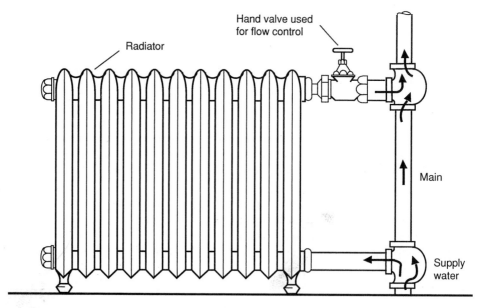

Figure 1-1 A hand valve is often used as a balancing valve on a radiator.

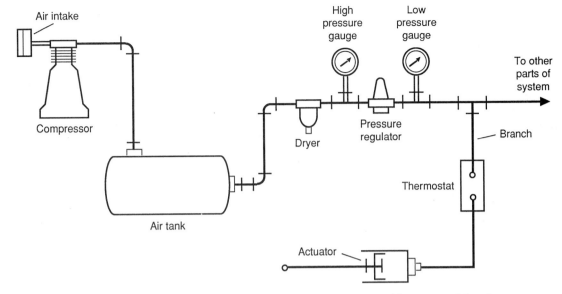

Figure 1-2 The parts of a pneumatic system are designed to provide a source of dry, clean air as a control signal.

or cooling. This was, and still is, a good control system, especially for control of hydronic systems.

The first centralized residential heating systems used warm air as a heating medium. In these warm-air heating systems air was heated as it went through a heat exchanger in a furnace and was then circulated by gravity through the building to be heated. While the first systems used gravity to move the air, later systems used blowers to move the air. Most of the first warm-air heating systems used solid fuel (wood or coal) furnaces as a heat source. The heat output of the furnace was a function of the amount of fuel burned and the amount of draft across the fire. Draft across the fire was controlled by dampers on the furnace. At first, regulation of the amount of fuel used and adjustment of the dampers was done manually. The first control systems developed for these furnaces still relied on manual control of the amount of fuel used, but the dampers were controlled by a control called a *thermostat*. The thermostat was located in a room above the furnace. A dial on the thermostat was connected to the damper doors on the furnace through a system of chains. If the thermostat was turned "up," the dampers opened to allow more draft across the fire, and consequently more heat. If the thermostat was turned "down," the dampers closed to allow less draft across the fire and less heat.

As gas- and oil-fired furnaces were developed, and their use common, firing was no longer controlled by dampers. Because of the manner in which the furnaces were fired, it was necessary to control heat output by cycling the burner off and on. This required more sophisticated controls than were used previously.. Electric controls were generally used to control these furnaces. A thermostat in the spaces being heated was used to turn the heating unit off and on in response to space temperature (Figure 1-3). A thermostat is an electric switch that is actuated by temperature. The thermostat controlled the flow of fuel to the unit burner, but draft across the fire was either fixed or self-adjusting.

VENTILATION CONTROL

The first use of ventilation for controlling the temperature in a building occurred long before recorded time. Opening doors and windows to let outside air in as ventilation air to keep the air in a building from getting stale is a basic form of ventilation. Using outside air to try to keep a building from getting too hot is a form of cooling by the use of ventilation. Both practices are as old as buildings themselves.

The design of many older buildings was dictated by the need to allow outside air for ventilation. Buildings with a large number of rooms were designed so that air could enter windows on one side of the building and cross-ventilate the building by leaving windows on the other side. Many older school buildings and public buildings were designed around this principle. Double-hung windows were also used so that they could be opened at both the top and bottom to encourage the circulation of outside air for ventilation. High-rise buildings were often built around a central well with the idea of using air from the well to ventilate the rooms on the inside of the building.

As fans were developed for circulating air, it became possible to bring outside air into a building's duct system through ventilation louvers and ducts. This is the system generally used today. The air is distributed to the spaces in the building through the heating or air-conditioning duct system. The air is also conditioned by the heating and air-conditioning equipment before it is

Figure 1-3 A thermostat is a temperature-actuated controller used in a typical control system

circulated. As these ducted ventilation systems are developed it is also important at certain times to control the flow of outside air. If the air outside is warmer than the air in a building, the minimum amount of outside air is wanted. But if the air outside is cooler than the air in a building, the outside air could be used to help cool the building. Dampers in the outside- and return-air ductwork make it possible to control the air, but manual control of the dampers is not too practical for this purpose (Figure 1-4).

Automatic control of outside- and return-air dampers is an ideal arrangement. A thermostat is installed in the mixed-air chamber to sense mixed-air temperature. The mixed-air chamber is the chamber where the outside air and return air come together. Another thermostat is installed in the outside air for monitoring air temperature. The thermostat in the mixed-air chamber controls the outside- and return-air dampers to provide a desired mixed-air temperature. The outside-air thermostat serves as a high limit to override the mixed-air thermostat when the outside air is too warm to be used for cooling purposes (Figure 1-5). This is the general control system still used to control ventilation air.

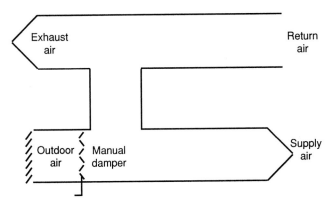

Figure 1-4 Ventilation for a typical building application is controlled by a damper.

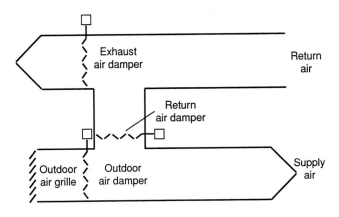

Figure 1-5 Outdoor air can be used to cool a building when the outdoor air is cool enough.

AIR-CONDITIONING CONTROL

Air-conditioning units and systems for cooling buildings were developed later than heating systems. The general use of air conditioning in buildings started early in the twentieth century. The control of air-conditioning units was then, and still is, generally achieved by cycling the compressor either by direct control from the thermostat or by a low-pressure limit control in the refrigerant system (Figure 1-6). The control of an air-conditioning system has not changed a great deal since systems for cooling by air conditioning were first introduced.

UNIT CONTROLS

Unit controls are the basic building blocks in a control system. They are the controls that operate the parts of each of the heating or air-conditioning units in the system (Figure 1-7). They are the thermostats, motor starters, valve

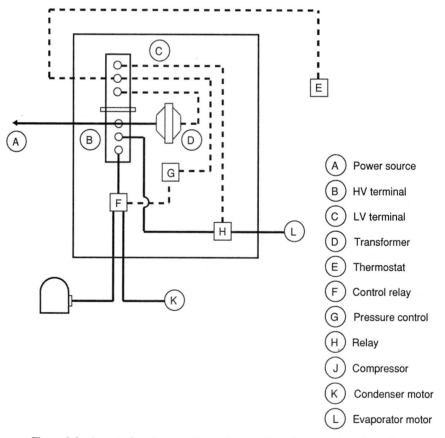

- (A) Power source
- (B) HV terminal
- (C) LV terminal
- (D) Transformer
- (E) Thermostat
- (F) Control relay
- (G) Pressure control
- (H) Relay
- (J) Compressor
- (K) Condenser motor
- (L) Evaporator motor

Figure 1-6 A control system can be used to regulate the operation of an air-conditioning system automatically.

```
Power controls
    Overcurrent protection
    Disconnects
        Fusing
        Circuit breakers
    Transformers

Operating controls
    Thermostat
    Relays
        Magnetic
        Contactors
        Magnetic starters
        Starting relays
    Solenoids
    Blower controls

Safety controls                  Heating only:
    Pressure switches                Combustion safety controls
        High pressure
        Low pressure
        Oil pressure
    Motor overloads
```

Figure 1-7 A control system can be divided into four subsystems, each containing controls for a specific purpose.

operators, and other controls that start and stop or otherwise control individual pieces of equipment. Unit controls can be categorized in one of four categories for heating systems and one of three for air-conditioning systems. In most heating systems there are electrical power controls, operating controls, safety controls, and combustion safety controls. Electric heating units do not have combustion safety controls, but they do have extra safety controls that provide the same functions as combustion safety controls. Air-conditioning units have electrical power controls, operating controls, and safety controls.

CONTROL SYSTEMS

A control system is the combination of unit controls and signal devices that tie those controls together (Figure 1-8). The control system is designed to help unit controls operate the individual pieces of equipment in the mechanical system in a way that will provide the level of comfort desired in a building. The main types of controls system used for comfort control at the present time are electrical, pneumatic, and various types of electronic systems. Each type of control system has operating characteristics that make it more useful in some applications than others. Electrical control systems are often used in smaller control applications, such as residences and small commercial buildings. Pneumatic controls lend themselves especially to control of hydronic heating and cooling systems in high-rise buildings. Electronic systems are fairly new, but they have applications that make them valuable for use in all systems. A control system may have electrical, pneumatic, and electronic controls in the same system (Figure 1-9).

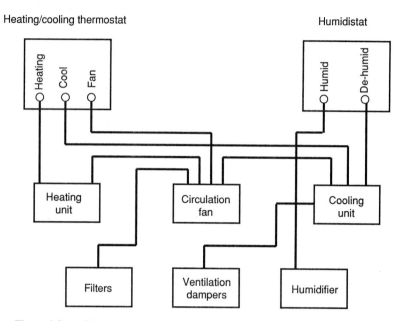

Figure 1-8 A block control system diagram shows the main components to be controlled along with the main control elements involved.

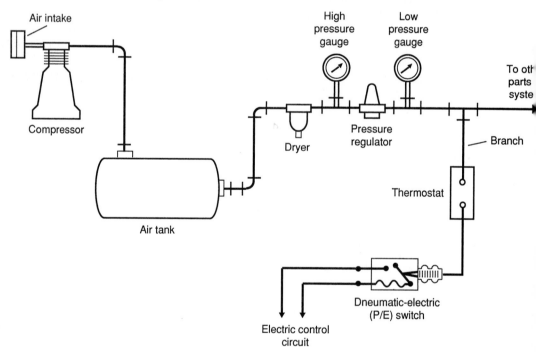

Figure 1-9 Through the use of transducers, more than one type of control system can be used in one control system.

TERMINOLOGY

In discussing controls and control systems, many specialized terms are used. A working knowledge of these terms is necessary for anyone working in the control field. At the end of the book is a short glossary listing some of the most commonly used control terms. Study of the terms will help a reader of this book, or any control text, better understand many of the terms commonly used in the control field.

One of the difficulties experienced in studying control systems is that controls that serve the same function are called by different names in different types of control systems. For instance, a control that senses temperature and sends a signal related to that temperature is called a *thermostat* in an electric control system, a *relay* in a pneumatic system, and a *controller* in an electronic control system. To try to avoid confusion, general terms will be used in this book when discussing control systems in general, but terms common to the type of control system will be used when discussing specific systems.

SUMMARY

The development of heating and cooling equipment and systems to provide comfortable conditions in our homes and commercial buildings made it necessary to develop control systems for the equipment. As the mechanical systems became more sophisticated, the control systems had to become more sophisticated. The main function of a control system is to operate the parts and components of the heating, ventilating, and air-conditioning systems to provide a comfortable atmosphere in a building.

The main types of control systems are electrical, pneumatic, and electronic. A control system may be made up of components of two or even all three types of systems. Each type of control system has special qualities that make it useful in certain applications.

QUESTIONS

1-1. What is the main function of an HVAC control system?

1-2. What two elements are necessary for control of a fuel-fired furnace?

1-3. Why is a control system that can only be adjusted for full-on or full-off operation not satisfactory?

1-4. What are the two main functions of ventilating air?

1-5. Ventilation air is usually circulated through a building through one of two types of duct systems. What are they?

1-6. What is the most common way of controlling air-conditioning systems?

1-7. What is the difference between control systems and unit controls?

1-8. Match the term in the first column with the phrase that best matches it in the second column by placing the letter that precedes the phrase in the space provided preceding the word.

A. _____ Power   a. provides safe operation
B. _____ Operating   b. turns on and off
C. _____ Safety   c. fires safely
D. _____ Combustion safe   d. controls electricity

1-9. Why do air-conditioning units have only three categories of controls instead of four as in a heating unit?

1-10. Name three types of control systems commonly used to control HVAC systems.

1-11. What is meant by the term *terminology?*

APPLICATION EXERCISES

1-1. On the attached diagram, draw in the pneumatic system piping in single lines:

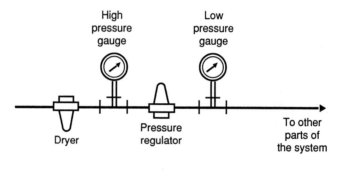

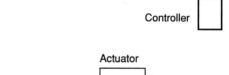

1-2. On the attached diagram, show the temperature at which a heating system would come on, and the temperature at which the cooling system would come on by arrows on the temperature line:

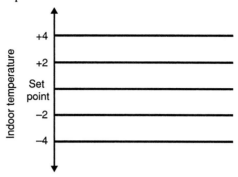

1-3. On the attached diagram show which control operates which damper motor, by drawing connecting lines.

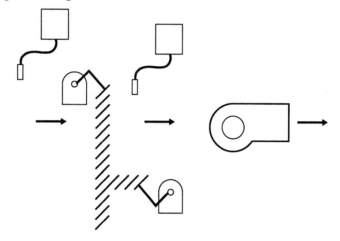

2

Comfort

Controls, as they relate to heating, air-conditioning, and ventilation systems, are those instruments and devices that control operation of the parts and components of the systems to provide a comfortable indoor climate in a building. The primary elements of a comfortable indoor climate are temperature, humidity, air movement, air cleanliness, and air freshness. The systems used to achieve a comfortable climate in a building are heating, ventilating, and air-conditioning systems.

INDOOR CLIMATE

Temperature

A comfortable indoor climate is one in which a person feels no sense of discomfort. Air temperature should be neither too warm nor too cold. The humidity level should be neither too dry nor too moist. The air should have enough movement to prevent temperature stratification and staleness; it should also be clean and contain enough ventilation or outdoor air to appear fresh.

Temperature control is an important element of a comfortable indoor climate system. To be comfortable, the temperature of the air in a building must be maintained at such a level that people in the building feel little or no difference between their own body temperature and that of the air around them. Most people feel comfortable at around 75°F in the summer, and because they generally wear more clothes in the winter, around 70°F in the winter. A true comfort control system maintains a comfortable temperature in a building both summer and winter.

Temperature is a measure of the intensity of heat. The temperature of the air is an indication of the amount of heat in the air. For the purposes of heating, air conditioning, and ventilation, temperature is measured in degrees Farenheit (°F) or degrees Celsius (°C). Temperature is measured with an instrument called a *thermometer* (Figure 2-1). There are many different types of thermometers available, from simple glass to very sophisticated electronic models. A glass thermometer is a glass tube filled with alcohol, mercury, or some other fluid with a low freezing-point temperature and a high boiling-point temperature. The tube has a thin glass bulb at the bottom and is sealed at the top. A temperature scale is etched on the glass tube. The temperature of the air surrounding the bulb determines the fluid level in the tube. If the temperature surrounding the bulb changes, the fluid in the glass bulb and tube expands or contracts. This causes the level of the fluid to rise or fall in the glass tube. The temperature scale on the glass tube is calibrated so that it will read the proper temperature for any level of the fluid. Other types of thermometers are available. Stem type thermometers indicate the temperature on a dial with a moving pointer. Electronic thermometers give a digital readout.

For heating applications temperature is usually measured in dry-bulb temperature only, but for air-conditioning applications both dry-bulb and wet-bulb temperatures are used (Figure 2-2). *Dry-bulb temperature* is a temperature measurement taken with a standard thermometer. It is a measurement of the intensity of *sensible heat,* the heat that can be sensed by a human being. *Wet-bulb temperature* is a measurement of temperature taken with a thermometer with a wet sock over the sensing bulb. The sock is saturated with water, and this cools the bulb somewhat by evaporation. A wet-bulb thermometer typically gives a lower reading than a dry-bulb thermometer. The rate of evaporation of the water from the sock on a wet-bulb thermometer is controlled by the amount of water vapor in the air. As a result, a wet-bulb reading is related to the amount of moisture or humidity in the air, and thus it indicates moisture content or latent heat in the air. *Latent heat* is heat related to a change of state but no change in sensible temperature. Water vapor exists in the air because of evaporation of water by natural causes. This is a change of state from liquid water to water vapor.

Humidity

Water vapor in the air is called *humidity.* We often think of the air surrounding us as made up of dry gases, while actually it nearly always contains some water vapor. Humidity is an important element in indoor climate control. Too much

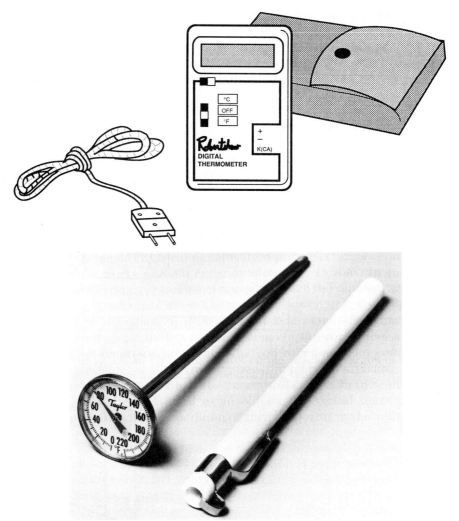

Figure 2-1 Thermometers are used to measure the intensity of heat at a given time and point.

or too little humidity makes a person feel uncomfortable. Humidity in the air is usually classified in one of two ways: as relative humidity or humidity ratio. *Relative humidity* is the percentage of water vapor in the air compared to the amount it would hold if it were saturated. *Humidity ratio* is the actual amount of water vapor in the air in pound (lb) or grains of moisture per pound of dry air. A grain is 1/7000 of a pound.

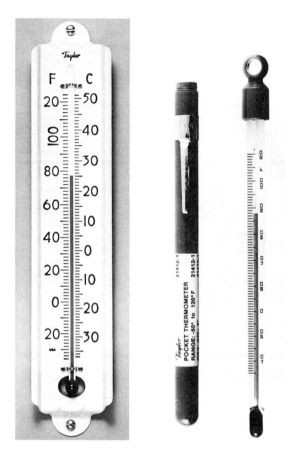

Figure 2-1 (continued)

Air Movement

Air movement is the motion of the air in a building measured in feet per minute (fpm). To be comfortable the air in a space should have constant, gentle air motion. Some air motion is required to prevent stagnation and stratification, but air motion that is too rapid feels drafty and cold. If the air in a building is completely still, it will quickly become stagnant. This is a condition in which oxygen in the air is depleted. In this condition the air is unpleasant to breathe. It will also quickly become dusty and may be contaminated with odors. *Stratification* is a condition in which air separates into different temperature levels (Figure 2-3). Typically, warmer air will rise to the ceiling and cooler air will lay along the floor. If there is no air motion, this condition can reach a point where the temperature layers are noticeable to occupants of the building in which they occur.

If air that is moving faster than about 40 to 50 fpm strikes a person, it feels drafty and cool. This is caused by increased evaporation from the skin of the person caused by the moving air. Below 40 to 50 fpm the moving air has little noticeable effect. To move the air in a building for ventilation or as part

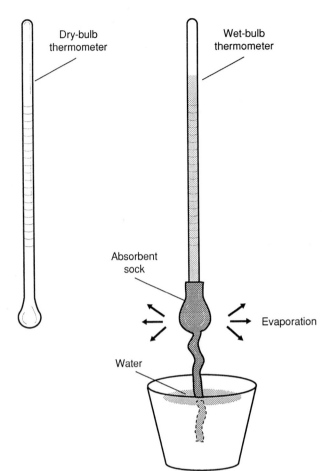

Figure 2-2 A dry-bulb thermometer is used to measure sensible heat, and a wet-bulb thermometer is used to measure latent heat.

of the heating and air-conditioning system, a mechanical device called a *fan* is used. A fan is a special blade used for moving air with its accompanying drives and a scroll. When the fan is enclosed in a cabinet, it is often called a *blower*. In a typical circulated-air system the system fan moves the air to the parts of a building through a duct system (Figure 2-4). For a heating or air-conditioning system the duct system includes a supply duct and registers and a return duct and grilles. For a ventilation system the fan is usually connected to a duct that brings air in from outside and other ducts that carry the air to spaces to be ventilated.

Clean air is a prime requirement for an indoor comfort system. The air in a building constantly picks up odors and vapors from the various processes going on in a building. As people move in and out of a building, they carry dirt and dust on their shoes or clothes. Perspiration also adds moisture and sometimes odor to the air. Cooking processes used in some buildings add odors and other contaminants to the air. Mechanical and chemical processes foul the

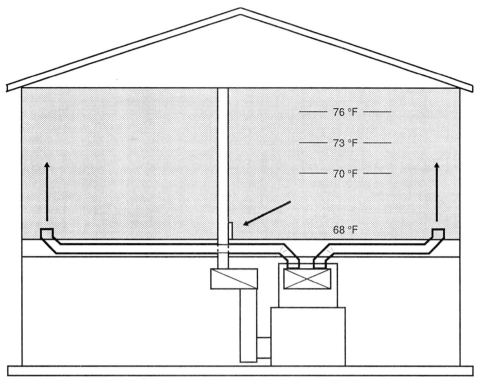

Figure 2-3 Stratification of air allows for different temperatures to exist in bands, usually warmer at the top and cooler at the bottom.

air. Finally, outgassing of chemical vapors from many materials used in construction and decoration in many buildings adds more contaminants. To keep a building comfortable, these contaminants must be removed or otherwise controlled.

Filters are located in the circulating air system in most buildings for cleaning the air in the buildings (Figure 2-5). These filters are installed in the return air ducts from a heating or air-conditioning system, or in the outdoor air duct in a ventilation system. Many different types of filters are used in circulating air systems. Basically, three types of filters are used. One type of filter commonly used in circulating air systems is that which screens particulate matter from the air. This type of filter is generally composed of fibrous or cellular material that is reasonably porous. The filter material is held in a frame and is placed in the duct through which air moves. This type of filter is generally designed to remove fairly large particulate matter from the air but is not highly efficient for particle sizes less than 10 micrometers (μm) in diameter.

A second type of filter material removes foreign matter by impingement or by catching it on the fibrous material. This type of filter usually has oil or some other viscous material on the fibers that holds the dirt particles. These filters are slightly more efficient than those listed above. Higher-efficiency filters

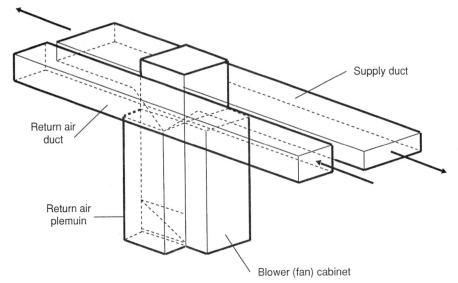

Figure 2-4 A blower together with a supply- and return-air duct make up a circulated-air system.

Supply duct

Return air duct

Return air plemuin

Blower (fan) cabinet

Figure 2-5 Filters of various kinds provide differing levels of filtration efficiency.

are available that are much less porous than fiber filters. Among these are those that use filter paper. They generally are reasonably efficient in removing particulate material down to about $1/10$ μm.

For the highest efficiency in filtration, special high-efficiency filters or electrostatic filtration is used. High-efficiency filters are either very thick or are in a bag type of configuration, to expose much filtering surface to the air. As air goes through an electrostatic filter, particulate matter in the air is ionized in one section of the filter and separated from the air by attraction to an electronic collector system.

Comfort   Chap. 2

Ventilation, the introduction of fresh air to a building, is an important element of an indoor comfort system. One of the best ways to prevent the buildup of unwanted odors or contaminants in a building is to dilute the indoor air with outside air. Although special filters can be used to remove some odors and contaminants, it is usually less expensive, and easier to control, to dilute them with ventilation air. Ventilation air is usually ducted from outside a building to the inside through a ventilation louver and duct. The louver is a grille located on the outside wall or roof of a building (Figure 2-6). It is built so that it will not admit rain or snow and is screened so that insects cannot get in. It should be located away from any exhaust air outlets or chimneys so that contaminated air will not enter. The ventilation air duct is a galvanized iron or fiberglass duct that runs from the ventilation louver to the fan used to distribute the air. If the air brought in for ventilation needs to be cleaned, filters are placed in the inlet duct ahead of the fan. Ventilation supply air ducts also run from the ventilation fan to the spaces in the building where the air is wanted. If the ventilation air is brought into a duct system on a heating or air-conditioning system it is brought in ahead of the filters in the system so that the air will be cleaned by the unit filters.

HEATING/COOLING MEDIA

A *heating or cooling medium* is the material that is heated or cooled by a furnace or air-conditioner and then used to carry the heat or cooling effect to a point of use. The two most commonly used media are air and water. Steam

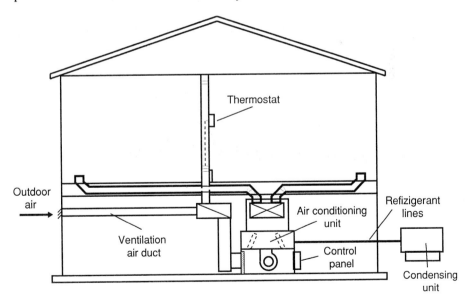

Figure 2-6 An outdoor grille and duct connected to a circulated-air return-air system allows ventilation air to be brought directly into a building.

Heating/Cooling Media

is used in some cases for heating-only applications. Although we usually think of the control medium as doing the actual heating or cooling in building spaces, it is only used to carry heat to a building or away from that building.

Air

Air is generally used as a medium when the point of use, or spaces to be conditioned, are not too far removed from the heating or cooling equipment (Figure 2-7). One of the chief advantages of using air as a medium is that it can be used directly; that is, it can be heated or cooled and then distributed directly into the spaces in a building. The air itself provides conditioning in the spaces.

Because air, as a gas, has a fairly high specific volume and fairly low specific heat, it requires fairly large ductwork for distribution. Specific volume is the space occupied by 1 lb of a material. Specific heat is the amount of heat required to raise the temperature of 1 lb of a material 1°F as compared to water. The specific volume of standard air is 13.28 ft³/lb. Standard air is dry air at a normal atmospheric pressure of 29.92 in. Hg and 68°F dry-bulb temperature. The specific heat of air is 0.24 Btu/lb. This means that air will carry 0.24 Btu per pound of air for each Farenheit degree of heat used to heat or cool it.

Water

Water is generally used as a medium when the heating or cooling equipment is located some distance from the spaces it is conditioning (Figure 2-8). When water is used as a medium it is heated or cooled in boilers or chillers. The conditioned water is then distributed to different points of use through a distribution system. In the spaces where heating or cooling is needed, the water is used to heat or cool air in terminal devices. The air is then used for heating or cooling the spaces. The water is simply a medium for carrying the heating or cooling effect.

Water is a better heating or cooling medium than air when heat has to be

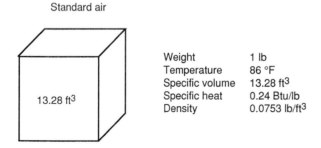

Standard air

Weight	1 lb
Temperature	86 °F
Specific volume	13.28 ft³
Specific heat	0.24 Btu/lb
Density	0.0753 lb/ft³

13.28 ft³

Figure 2-7 The availability of air makes it valuable as a carrier of heat.

Comfort   Chap. 2

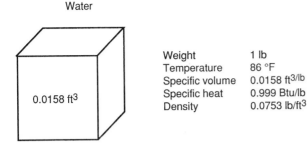

Water

Weight	1 lb
Temperature	86 °F
Specific volume	0.0158 ft³/lb
Specific heat	0.999 Btu/lb
Density	0.0753 lb/ft³

0.0158 ft³

Figure 2-8 The thermodynamic properties of water make it an excellent carrier of heat.

moved some distance because of its smaller specific volume and its greater ability to hold heat. The specific volume of water is 0.0160 ft³/lb, and its specific heat is 1.00. One pound of water occupies only 0.0160 ft³ of space, and it will carry 1.00 Btu of heat per pound for each degree of heat used to heat or cool it. These characteristics make it possible to use much smaller pipe to transport heating or cooling effects with water than with an air duct.

CONTROL POINTS

Control points are points at which sensors are located to sense conditions to be controlled by a control system. In most cases the conditions sensed are relative to air conditions, so the control points are located to help sense those conditions. There are three general locations in or around a building at which control points are located: inside the space to be conditioned, outside the space in the outside air, or in the air or water used as a medium (Figure 2-9). A combination of any of the above may be used.

Control points located in the conditioned space are normally used in applications where the major factors affecting the space conditions are located in the space and at the same time, fairly close control of the conditions is required. An example is a building with high internal gain because of people, but requiring close control of the temperature and humidity for comfort. When a control point is located inside a space to be conditioned, it should be located where it will sense average conditions (Figure 2-10). A thermostat or humidistat should be located on an inside partition about 5 feet from the floor. The control should always be located where it senses indoor conditions only and where outside influences such as sunlight and drafts will not affect the conditions sensed.

Outside control points are located outside the building for which they are providing control. They are usually used for applications in which outside ambient conditions affect the indoor climate significantly. Usually, outside control points do not provide close control of indoor conditions but are often used as outdoor resets for indoor controls. An *outdoor reset* is a control that resets the indoor set point in response to changes in outdoor conditions. For instance, if the temperature outdoors goes down, the set-point temperature in the building goes up.

Control Points

1. Inside conditions
    Residence
    Commerical
        Excellent control when conditions can be sensed
        from a central location

2. Outside conditions
    Commerical
        Used for resetting indoor conditions to offset
        changing climate conditions, and for determining
        startup time after a setback
    Industrial
        Reset of medium conditions

3. Medium conditions
    Commerical
    Industrial
        Return-air conditions:
        goods for sensing changing loads
        Supply-air conditions:
        good for controlling discharge-air conditions

Figure 2-9 Three basic locations for controllers are inside a building, outside, and in the conditioned medium.

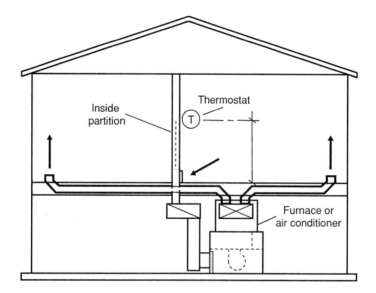

Figure 2-10 A room thermostat should be located where it will sense average air conditions in a building.

Control points used for sensing solar radiation are one example of outside control points. Another example is the use of outdoor thermostats used as low-temperature controllers for electric heating units. When control is by means of sensing the condition of the control agent, the control sensor is located in the medium that is being controlled, usually air or water. One type of medium control is having the sensor located in the supply medium as it leaves the heating or air-conditioning equipment (Figure 2-11). By controlling the temperature of the supply air or water that is used to condition a space, close control of space conditions can be maintained.

A variation of medium control is the use of a sensor in the return air or water from the conditioned spaces in a building. Although this closely approximates space control, it is also a form of medium control. Control of space conditions can be maintained in situations where rapidly changing conditions occur by using return air control with carefully sized heating and air-conditioning equipment.

Combinations of any of the control point locations described above may be used for specific applictions. Outdoor reset may be used with either space control or medium control (Figure 2-12). These combinations are often used when changing outdoor conditions cause variations in indoor conditions more rapidly than the space controller can sense or react to. Medium control is often used in combination with space control when discharge-air conditions need to be monitored carefully. In such applications the medium controller provides a limit function.

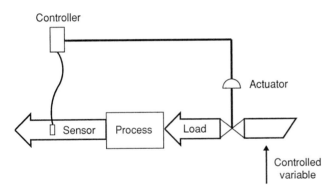

Figure 2-11 A thermostat located in the supply-air medium provides a limit control function.

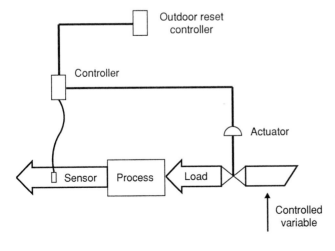

Figure 2-12 Outdoor reset used with an indoor or medium sensor provides excellent control.

SUMMARY

Control systems are used to tie together the various components and parts of a heating, air-conditioning, or ventilations system and to operate and regulate system operation. A system should be operated to provide a comfortable indoor climate in the building in which it is installed.

Two basic control agents or media are used to control the climate in a building: air and water. The media are used to carry the heating or cooling effect to the various parts of the building.

Three control points are used for locating sensors in a control system: in the building, outside the building, or in the control agent or medium used for conditioning the building. Any combination of control points may also be used.

QUESTIONS

2-1. Name four elements of a comfortable climate.

2-2. At what temperature do most people feel comfortable in a building in the winter?

2-3. Define the term *sensible heat*.

2-4. Define the term *latent heat*.

2-5. *True or false:* Water vapor in the air represents latent heat.

2-6. Name two ways in which humidity (water vapor in the air) is classified.

2-7. Define *relative humidity*.

2-8. What type of thermometer is used to determine how much moisture there is in the air?

2-9. Why is air circulation in a building necessary for a comfortable climate?

2-10. *True or false:* When the air in a building becomes temperature stratified, the temperature at the ceiling is usually lower than that at the floor.

2-11. Filters are used to _____(clean/humidify/heat) the air as it circulates through the duct system in a building.

2-12. What two types of filters are used to achieve the highest degree of filtration of the air in a building?

2-13. Ventilation air is used to prevent the buildup of unwanted _____ or _____in a building.

2-14. Name the two most commonly used media for carrying heat from point of origin to point of use.

2-15. Why is water considered a better medium than air for carrying heat?

2-16. What two characteristics can be compared in deciding which material will carry heat well?

2-17. Name three control points that can be used for sensing conditions for a control system.

2-18. What is an outdoor reset control system?

APPLICATION EXERCISES

2-1. Using a standard dry-bulb thermometer, measure the air temperature outside your classroom in the sunshine and in the shade; and inside at several different locations, such as classroom, library, or cafeteria. Record the temperature at each location, and show the difference between each temperature and the classroom temperature. Explain what causes the difference.

		Temperature	Difference from base
Outside	Shade	_____	_____
	Sunlight	_____	_____
Inside	Classroom	_____	_____
	Library	_____	_____
	Cafeteria	_____	_____
	Other	_____	_____

2-2. Using a standard psychrometer with both dry-bulb and wet-bulb thermometers, measure both dry-bulb and wet-bulb temperatures outside your classroom, in the sunshine and in the shade; and inside at several different locations, such as classroom, library, or cafeteria. Record both temperatures at each location, and show the difference between each temperature and the classroom temperature as a base. Explain why the difference exists.

		Temperature	Difference from base
Outside	Shade	_____	_____
	Sunlight	_____	_____
Inside	Classroom	_____	_____
	Library	_____	_____
	Cafeteria	_____	_____
	Other	_____	_____

2-3. The dry air in a room weighs 115.23 lbs. At a humidity ratio of 60 gr/lb how many lbs of water vapor is in the air in the room?

2-4. How much heat, in Btu, is required to raise the temperature of 938 cu ft of air at standard conditions 25° F?

2-5. How much heat is required, in Btu, to heat 100 lbs of water 25° F?

3

Control Theory

The function of an HVAC control system is to operate the heating, ventilating, and air-conditioning equipment in a building in such a way as to provide a comfortable atmosphere in that building. The control system must be matched with the type of equipment used in order to operate the equipment properly.

EQUIPMENT OPERATION

Heating and air-conditioning units produce heat or cooling at a specified rate. This is measured in Btu per hour (Btu/h), Btu (British thermal unit) being a measure of heat content. A furnace with an output rating of 100,000 Btu/h produces heat at a rate of 100,000 Btu per hour. An air conditioner with a cooling capacity of 36,000 Btu/h produces a cooling effect at the rate of 36,000 Btu per hour.

The heating or air-conditioning equipment chosen for each building must have the output capacity to heat or cool that building to a selected indoor temperature at any outdoor temperature. The inside design temperature should be one that people find most comfortable. This is usually 70°F for heating and 75°F for cooling, and the outside temperature should be any temperature down to a reasonable temperature for the area in which the building is constructed (Figure 3-1).

The *outside design temperature* is the coldest temperature for winter and the warmest for summer that is expected in an area, disregarding those coldest and warmest temperatures that occur 2½% of the time (Figure 3-2). Outdoor design temperatures are available in tables prepared from records kept by the U.S. Weather Service. Since heating and air-conditioning equipment is selected to heat or cool a building during the worst possible conditions, and those conditions occur only occasionally, the equipment actually is oversized for the heating or cooling requirements that exist most of the time. If the equipment runs 100% of the time, too much heat or cooling is being provided for the space most of the time. To solve this problem the equipment is run in one of three ways (Figure 3-3). The first is to turn it on and off in relatively short cycles. This is called *digital operation.* The second is to bring the equipment on in incremental stages, each increment being a percentage of the total output, until the full output is on. This is called *stepped,* or *sequenced, operation.* The third is to modulate the output of the equipment so that the output matches the heating or cooling load at any given time. This is called *proportional operation.*

Each type of heating or air-conditioning unit functions in such a way that it falls into one of the three operational modes listed above. Most small resi-

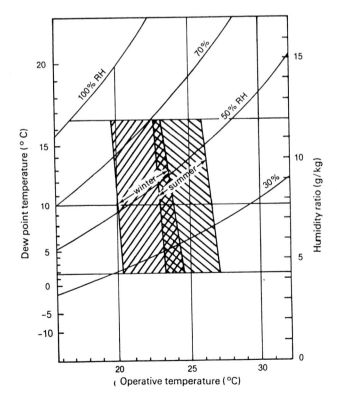

Figure 3-1 Most people feel comfortable at around 70°F and 50% relative humidity.

TABLE 1 CLIMATIC CONDITIONS FOR THE UNITED STATES

Col. 1	Col. 2		Col. 3		Col. 4	Winter,[b] °F Col. 5		Summer,[c] °F Col. 6			Col. 7	Col. 8			Prevailing Wind Col. 9		Temp. °F Col. 10	
State and Station[a]	Lat.		Long.		Elev.	Design Dry-Bulb		Design Dry-Bulb and Mean Coincident Wet-Bulb			Mean Daily	Design Wet-Bulb			Winter	Summer	Median of Annual Extr.	
	°	′	°	′	Feet	99% 97.5%		1%	2.5%	5%	Range	1% 2.5% 5%			Knots[d]		Max.	Min.
ALABAMA																		
Alexander City	32	57	85	57	660	18 22		96/77	93/76	91/76	21	79	78	78				
Anniston AP	33	35	85	51	599	18 22		97/77	94/76	92/76	21	79	78	78	SW 5	SW	98.4	12.4
Auburn	32	36	85	30	652	18 22		96/77	93/76	91/76	21	79	78	78			99.8	14.6
Birmingham AP	33	34	86	45	620	17 21		96/74	94/75	92/74	21	78	77	76	NNW 8	WNW	98.5	12.9
Decatur	34	37	86	59	580	11 16		95/75	93/74	91/74	22	78	77	76				
Dothan AP	31	19	85	27	374	23 27		94/76	92/76	91/76	20	80	79	78				
Florence AP	34	48	87	40	581	17 21		97/74	94/74	92/74	22	78	77	76	NW 7	NW		
Gadsden	34	01	86	00	554	16 20		96/75	94/75	92/74	22	78	77	76	NNW 8	WNW		
Huntsville AP	34	42	86	35	606	11 16		95/75	93/74	91/74	23	78	77	76	N 9	SW		
Mobile AP	30	41	88	15	211	25 29		95/77	93/77	91/76	18	80	79	78	N 10	N		
Mobile Co	30	40	88	15	211	25 29		95/77	93/77	91/76	16	80	79	78			97.9	22.3
Montgomery AP	32	23	86	22	169	22 25		96/76	95/76	93/76	21	79	79	78	NW 7	W	98.9	18.2
Selma-Craig AFB	32	20	87	59	166	22 26		97/78	95/77	93/77	21	81	80	79	N 9	SW	100.1	17.6
Talladega	33	27	86	06	565	18 22		97/77	94/76	92/76	21	79	78	78			99.6	11.2
Tuscaloosa AP	33	13	87	37	169	20 23		98/75	96/76	94/76	22	79	78	77	N 5	WNW		
ALASKA																		
Anchorage AP	61	10	150	01	114	−23 −18		71/59	68/58	66/56	15	60	59	57	SE 3	WNW		
Barrow (S)	71	18	156	47	31	−45 −41		57/53	53/50	49/47	12	54	50	47	SW 8	SE		
Fairbanks AP (S)	64	49	147	52	436	−51 −47		82/62	78/60	75/59	24	64	62	60	N 5	S		
Juneau AP	58	22	134	35	12	−4 1		74/60	70/58	67/57	15	61	59	58	N 7	W		
Kodiak	57	45	152	29	73	10 13		69/58	65/56	62/55	10	60	58	56	WNW 14	NW		
Nome AP	64	30	165	26	13	−31 −27		66/57	62/55	59/54	10	58	56	55	N 4	W		
ARIZONA																		
Douglas AP	31	27	109	36	4098	27 31		98/63	95/63	93/63	31	70	69	68			104.4	14.0
Flagstaff AP	35	08	111	40	7006	−2 4		84/55	82/55	80/54	31	61	60	59	NE 5	SW	90.0	−11.6
Fort Huachuca AP (S)	31	35	110	20	4664	24 28		95/62	92/62	90/62	27	69	68	67	SW 5	W		
Kingman AP	35	12	114	01	3539	18 25		103/65	100/64	97/64	30	70	69	69				
Nogales	31	21	110	55	3800	28 32		99/64	96/64	94/64	31	71	70	69	SW 5	W		
Phoenix AP (S)	33	26	112	01	1112	31 34		109/71	107/71	105/71	27	76	75	75	E 4	W	112.8	26.7
Prescott AP	34	39	112	26	5010	4 9		96/61	94/60	92/60	30	66	65	64				
Tucson AP (S)	32	07	110	56	2558	28 32		104/66	102/66	100/66	26	72	71	71	SE 6	WNW	108.9	.3
Winslow AP	35	01	110	44	4895	5 10		97/61	95/60	93/60	32	66	65	64	SW 6	WSW	102.7	−.4
Yuma AP	32	39	114	37	213	36 39		111/72	109/72	107/71	27	79	78	77	NNE 6	WSW	114.8	30.8
ARKANSAS																		
Blytheville AFB	35	57	89	57	264	10 15		96/78	94/77	91/76	21	81	80	78	N 8	SSW		
Camden	33	36	92	49	116	18 23		98/76	96/76	94/76	21	80	79	78				
El Dorado AP	33	13	92	49	277	18 23		98/76	96/76	94/76	21	80	79	78	S 6	SE	101.0	13.9
Fayetteville AP	36	00	94	10	1251	7 12		97/72	94/73	92/73	23	77	76	75	NE 9	SSW	99.4	−.3
Fort Smith AP	35	20	94	22	463	12 17		101/75	98/76	95/76	24	80	79	78	NW 8	SW	101.9	7.0
Hot Springs	34	29	93	06	535	17 23		101/77	97/77	94/77	22	80	79	78	N 8	SW	103.0	10.6
Jonesboro	35	50	90	42	345	10 15		96/78	94/77	91/76	21	81	80	78			101.7	7.3
Little Rock AP (S)	34	44	92	14	257	15 20		99/76	96/77	94/77	22	80	79	78	N 9	SSW	99.0	11.2
Pine Bluff AP	34	18	92	05	241	16 22		100/78	97/77	95/78	22	81	80	80	N 7	SW	102.2	13.1
Texarkana AP	33	27	93	59	389	18 23		98/76	96/77	93/76	21	80	79	78	WNW 9	SSW	104.8	14.0

[a] AP or AFB following the station name designates airport or Airforce base temperature observations. Co designates office locations within an urban area that re affected by the surrounding area. Undesignated stations are semirural and may compared to airport data.

[b] Winter design data are based on the 3-month period, December through February.
[c] Summer design data are based on the 4-month period, June through September.
[d] Mean wind speeds occurring coincidentally with the 99.5% dry-bulb winter design temperature.

Figure 3-2 Outside design temperatures can be found in tables prepared by ASHRAE or other organizations concerned with indoor comfort.

dential furnaces that burn fuels are of the on–off type and are best controlled by digital control. Because of the way they are built, electric furnaces can easily be staged and are best controlled by step control. Hydronic heating systems that use hot water or steam coils for producing heat are easily operated as modulating units and are best controlled by proportional control. Most air-conditioning units operate basically as on–off units and are operated digitally or by step controls.

TABLE 1 CLIMATIC CONDITIONS FOR THE UNITED STATES (*Continued*)

State and Station[a]	Lat. °	′	Long. °	′	Elev. Feet	Winter Design Dry-Bulb 99%	97.5%	Summer Design Dry-Bulb and Mean Coincident Wet-Bulb 1%	2.5%	5%	Mean Daily Range	Design Wet-Bulb 1%	2.5%	5%	Prevailing Wind Winter (Knots)		Summer	Median of Annual Extr. Max.	Min.
CALIFORNIA																			
Bakersfield AP	35	25	119	03	475	30	32	104/70	101/69	98/68	32	73	71	70	ENE	5	WNW	109.8	25.3
Barstow AP	34	51	116	47	1927	26	29	106/68	104/68	102/67	37	73	71	70	WNW	7	W	110.4	17.4
Blythe AP	33	37	114	43	395	30	33	112/71	110/71	108/70	28	75	75	74				116.8	24.1
Burbank AP	34	12	118	21	775	37	39	95/68	91/68	88/67	25	71	70	69	NW	3	S		
Chico	39	48	121	51	238	28	30	103/69	101/68	98/67	36	71	70	68	NW	5	SSE	109.0	22.6
Concord	37	58	121	59	200	24	27	100/69	97/68	94/67	32	71	70	68	WNW	5	NW		
Covina	34	05	117	52	575	32	35	98/69	95/68	92/67	31	73	71	70					
Crescent City AP	41	46	124	12	40	31	33	68/60	65/59	63/58	18	62	60	59					
Downey	33	56	118	08	116	37	40	93/70	89/70	86/69	22	72	71	70					
El Cajon	32	49	116	58	367	42	44	83/69	80/69	78/68	30	71	70	68					
El Centro AP (S)	32	49	115	40	− 43	35	38	112/74	110/74	108/74	34	81	80	78	W	6	SE		
Escondido	33	07	117	05	660	39	41	89/68	85/68	82/68	30	71	70	69					
Eureka/Arcata AP	40	59	124	06	218	31	33	68/60	65/59	63/58	11	62	60	59	E	5	NW	75.8	29.7
Fairfield-Travis AFB	38	16	121	56	62	29	32	99/68	95/67	91/66	34	70	68	67	N	5	WSW		
Fresno AP (S)	36	46	119	43	328	28	30	102/70	100/69	97/68	34	72	71	70	E	4	WNW	108.7	25.8
Hamilton AFB	38	04	122	30	3	30	32	89/68	84/66	80/65	28	72	69	67	N	4	SE		
Laguna Beach	33	33	117	47	35	41	43	83/68	80/68	77/67	18	70	69	68					
Livermore	37	42	121	57	545	24	27	100/69	97/68	93/67	24	71	70	68	WNW	4	NW		
Lompoc, Vandenberg AFB	34	43	120	34	368	35	38	75/61	70/61	67/60	20	63	61	60	ESE	5	NW		
Long Beach AP	33	49	118	09	30	41	43	83/68	80/68	77/67	22	70	69	68	NW	4	WNW		
Los Angeles AP (S)	33	56	118	24	97	41	43	83/68	80/68	77/67	15	70	69	68	E	4	WSW		
Los Angeles Co (S)	34	03	118	14	270	37	40	93/70	89/70	86/69	20	72	71	70	NW	4	NW	98.1	35.9
Merced-castle AFB	37	23	120	34	188	29	31	102/70	99/69	96/68	36	72	71	70	ESE	4	NW		
Modesto	37	39	121	00	91	28	30	101/69	98/68	95/67	36	71	70	69				105.8	26.2
Monterey	36	36	121	54	39	35	38	75/61	71/61	68/61	20	64	62	61	SE	4	NW		
Napa	38	13	122	17	56	30	32	100/69	96/68	92/67	30	71	69	68				103.1	25.8
Needles AP	34	46	114	37	913	30	33	112/71	110/71	108/70	27	75	75	74				116.4	26.7
Oakland AP	37	49	122	19	5	34	36	85/64	80/63	75/62	19	66	64	63	E	5	WNW	93.0	31.8
Oceanside	33	14	117	25	26	41	43	83/68	80/68	77/67	13	70	69	68					
Ontario	34	03	117	36	952	31	33	102/70	99/69	96/67	36	74	72	71	E	4	WSW		
Oxnard	34	12	119	11	49	34	36	83/66	80/64	77/63	19	70	68	67					
Palmdale AP	34	38	118	06	2542	18	22	103/65	101/65	98/64	35	69	67	66	SW	5	WSW		
Palm Springs	33	49	116	32	411	33	35	112/71	110/70	108/70	35	76	74	73					
Pasadena	34	09	118	09	864	32	35	98/69	95/68	92/67	29	73	71	70				102.8	30.4
Petaluma	38	14	122	38	16	26	29	94/68	90/66	87/65	31	72	70	68				102.0	24.2
Pomona Co	34	03	117	45	934	28	30	102/70	99/69	95/68	36	74	72	71	E	4	W	105.7	26.2
Redding AP	40	31	122	18	495	29	31	105/68	102/67	100/66	32	71	69	68				109.2	26.0
Redlands	34	03	117	11	1318	31	33	102/70	99/69	96/68	33	74	72	71				106.7	27.1
Richmond	37	56	122	21	55	34	36	85/64	80/63	75/62	17	66	64	63					
Riverside-March AFB (S)	33	54	117	15	1532	29	32	100/68	98/68	95/67	37	72	71	70	N	4	NW	107.6	26.6
Sacramento AP	38	31	121	30	17	30	32	101/70	98/70	94/69	36	72	71	70	NNW	6	SW	105.1	27.6
Salinas AP	36	40	121	36	75	30	32	74/61	70/60	67/59	24	62	61	59					
San Bernardino, Norton AFB	34	08	117	16	1125	31	33	102/70	99/69	96/68	38	74	72	71	E	3	NW	109.3	25.3
San Diego AP	32	44	117	10	13	42	44	83/69	80/69	78/68	12	71	70	68	NE	3	WNW	91.2	37.4
San Fernando	34	17	118	28	965	37	39	95/68	91/68	88/67	38	71	70	69					
San Francisco AP	37	37	122	23	8	35	38	82/64	77/63	73/62	20	65	64	62	S	5	NW		
San Francisco Co	37	46	122	26	72	38	40	74/63	71/62	69/61	14	64	62	61	W	5	W	91.3	35.9
San Jose AP	37	22	121	56	56	34	36	85/66	81/65	77/64	26	68	67	65	SE	4	NNW	98.6	28.2
San Luis Obispo	35	20	120	43	250	33	35	92/69	88/70	84/69	26	73	71	70	E	4	W	99.8	29.3
Santa Ana AP	33	45	117	52	115	37	39	89/69	85/68	82/68	28	71	70	69	E	3	SW	101.0	29.9
Santa Barbara MAP	34	26	119	50	10	34	36	81/67	77/66	75/65	24	68	67	66	NE	3	SW	97.1	31.7
Santa Cruz	36	59	122	01	125	35	38	75/63	71/61	68/61	28	64	62	61				97.5	26.8
Santa Maria AP (S)	34	54	120	27	236	31	33	81/64	76/63	73/62	23	65	64	63	E	4	WNW		
Santa Monica Co	34	01	118	29	64	41	43	83/68	80/68	77/67	16	70	69	68					
Santa Paula	34	21	119	05	263	33	35	90/68	86/67	84/66	36	71	69	68					
Santa Rosa	38	31	122	49	125	27	29	99/68	95/67	91/66	34	70	68	67	N	5	SE	102.5	23.4
Stockton AP	37	54	121	15	22	28	30	100/69	97/68	94/67	37	71	70	68	WNW	4	NW	104.1	24.5
Ukiah	39	09	123	12	623	27	29	99/69	95/68	91/67	40	70	68	67				108.1	21.6
Visalia	36	20	119	18	325	28	30	102/70	100/69	97/68	38	72	71	70				108.4	25.1
Yreka	41	43	122	38	2625	13	17	95/65	92/64	89/64	38	67	65	64				102.8	7.1
Yuba City	39	08	121	36	80	29	31	104/68	101/67	99/66	36	71	69	68					
COLORADO																			
Alamosa AP	37	27	105	52	7537	− 21	− 16	84/57	82/57	80/57	35	62	61	60				96.0	− 8.4
Boulder	40	00	105	16	5445	2	8	93/59	91/59	89/59	27	64	63	62					
Colorado Springs AP	38	49	104	43	6145	− 3	2	91/58	88/57	86/57	30	63	62	61	N	9	S	92.3	− 12.1
Denver AP	39	45	104	52	5283	− 5	1	93/59	91/59	89/59	28	64	63	62	S	8	SE	96.8	− 10.4
Durango	37	17	107	53	6550	− 1	4	89/59	87/59	85/59	30	64	63	62				92.4	− 11.2
Fort Collins	40	35	105	05	4999	− 10	− 4	93/59	91/59	89/59	28	64	63	62				95.2	− 18.1
Grand Junction AP (S)	39	07	108	32	4843	2	7	96/59	94/59	92/59	29	64	63	62	ESE	5	WNW	99.9	− 3.4
Greeley	40	26	104	38	4648	− 11	− 5	96/60	94/60	92/60	29	65	64	63					
Lajunta AP	38	03	103	30	4160	− 3	3	100/68	98/68	95/67	31	72	70	69	W	8	S		
Leadville	39	15	106	18	10155	− 8	− 4	84/51	81/51	78/50	30	56	55	54				79.7	− 17.8
Pueblo AP	38	18	104	29	4641	− 7	0	97/61	95/61	92/61	31	67	66	65	W	5	SE	100.5	− 12.2
Sterling	40	37	103	12	3939	− 7	− 2	95/62	93/62	90/62	30	67	66	65				100.3	− 15.4
Trinidad AP	37	15	104	20	5740	− 2	3	93/61	91/61	89/61	32	66	65	64	W	7	WSW	96.8	− 10.5

TABLE 1   CLIMATIC CONDITIONS FOR THE UNITED STATES (*Continued*)

Col. 1	Col. 2		Col. 3		Col. 4	Col. 5 Winter,[b] °F Design Dry-Bulb		Col. 6 Summer,[c] °F Design Dry-Bulb and Mean Coincident Wet-Bulb			Col. 7 Mean Daily	Col. 8 Design Wet-Bulb			Col. 9 Prevailing Wind		Col. 10 Temp. °F Median of Annual Extr.	
State and Station[a]	Lat. ° '		Long. ° '		Elev. Feet	99%	97.5%	Mean 1%	Coincident 2.5%	5%	Range	1%	2.5%	5%	Winter Knots[d]	Summer	Max.	Min.
ILLINOIS																		
Aurora	41	45	88	20	744	−6	−1	93/76	91/76	88/75	20	79	78	76			96.7	−13.0
Belleville, Scott AFB	38	33	89	51	453	1	6	94/76	92/76	89/75	21	79	78	76	WNW 8	S		
Bloomington	40	29	88	57	876	−6	−2	92/75	90/74	88/73	21	78	76	75			98.4	−9.6
Carbondale	37	47	89	15	417	2	7	95/77	93/77	90/76	21	80	79	77			100.9	−.8
Champaign/Urbana	40	02	88	17	777	−3	2	95/75	92/74	90/73	21	78	77	75				
Chicago, Midway AP	41	47	87	45	607	−5	0	94/74	91/73	88/72	20	77	75	74	NW 11	SW		
Chicago, O'Hare AP	41	59	87	54	658	−8	−4	91/74	89/74	86/72	20	77	76	74	WNW 9	SW		
Chicago Co	41	53	87	38	590	−3	2	94/75	91/74	88/73	15	79	77	75				
Danville	40	12	87	36	695	−4	1	93/75	90/74	88/73	21	78	77	75			96.1	−8.3
Decatur	39	50	88	52	679	−3	2	94/75	91/74	88/73	21	78	77	75	W 10	SSW	98.2	−8.4
Dixon	41	50	89	29	696	−7	−2	93/75	90/74	88/73	23	78	77	75	NW 10	SW	99.0	−8.1
Elgin	42	02	88	16	758	−7	−2	91/75	88/74	86/73	21	78	77	75			97.5	−13.5
Freeport	42	18	89	37	780	−9	−4	91/74	89/73	87/72	24	77	76	74				
Galesburg	40	56	90	26	764	−7	−2	93/75	91/75	88/74	22	78	77	75	WNW 8	SW		
Greenville	38	53	89	24	563	−1	4	94/76	92/75	89/74	21	79	78	76				
Joliet	41	31	88	10	582	−5	0	93/75	90/74	88/73	20	78	77	75	NW 11	SW		
Kankakee	41	05	87	55	625	−4	1	93/75	90/74	88/73	21	78	77	75				
La Salle/Peru	41	19	89	06	520	−7	−2	93/75	91/75	88/74	22	78	77	75				
Macomb	40	28	90	40	702	−5	0	95/76	92/76	89/75	22	79	78	76				
Moline AP	41	27	90	31	582	−9	−4	93/75	91/75	88/74	23	78	77	75	WNW 8	SW	96.8	−12.7
Mt Vernon	38	19	88	52	479	0	5	95/76	92/75	89/74	21	79	78	76			100.5	−2.9
Peoria AP	40	40	89	41	652	−8	−4	91/75	89/74	87/73	22	78	76	75	WNW 8	SW	98.0	−10.9
Quincy AP	39	57	91	12	769	−2	3	96/76	93/76	90/76	22	80	78	77	NW 11	SSW	101.1	−6.7
Rantoul, Chanute AFB	40	18	88	08	753	−4	1	94/75	91/74	89/73	21	78	77	75	W 10	SSW		
Rockford	42	21	89	03	741	−9	−4	91/74	89/73	87/72	24	77	76	74			97.4	−13.8
Springfield AP	39	50	89	40	588	−3	2	94/75	92/74	89/74	21	79	77	76	NW 10	SW	98.1	−7.2
Waukegan	42	21	87	53	700	−6	−3	92/76	89/74	87/73	21	78	76	75			96.5	−10.6
INDIANA																		
Anderson	40	06	85	37	919	0	6	95/76	92/75	89/74	22	79	78	76	W 9	SW	95.1	−6.0
Bedford	38	51	86	30	670	0	5	95/76	92/75	89/74	22	79	78	76			97.5	−4.4
Bloomington	39	08	86	37	847	0	5	95/76	92/75	89/74	22	79	78	76	W 9	SW	97.8	−4.6
Columbus, Bakalar AFB	39	16	85	54	651	3	7	95/76	92/75	90/74	22	79	78	76	W 9	SW	98.3	−6.4
Crawfordsville	40	03	86	54	679	−2	3	94/75	91/74	88/73	22	79	77	76			98.4	−7.6
Evansville AP	38	03	87	32	381	4	9	95/76	93/75	91/75	22	79	78	77	NW 9	SW	98.2	.2
Fort Wayne AP	41	00	85	12	791	−4	1	92/73	89/72	87/72	24	77	75	74	WSW 10	SW		
Goshen AP	41	32	85	48	827	−3	1	91/73	89/73	86/72	23	77	75	74			96.8	−10.5
Hobart	41	32	87	15	600	−4	2	91/73	88/73	85/72	21	77	75	74			98.5	−8.5
Huntington	40	53	85	30	802	−4	1	92/73	89/72	87/72	23	77	75	74			96.9	−8.1
Indianapolis AP	39	44	86	17	792	−2	2	92/74	90/74	87/73	22	78	76	75	WNW 10	SW	96	−7
Jeffersonville	38	17	85	45	455	5	10	95/74	93/74	90/74	23	79	77	76			98	2
Kokomo	40	25	86	03	855	−4	0	91/74	90/73	88/73	22	77	75	74			98.2	−7.5
Lafayette	40	2	86	5	600	−3	3	94/74	91/73	88/73	22	78	76	75				
La Porte	41	36	86	43	810	−3	3	93/74	90/74	87/73	22	78	76	75			98.1	−10.5
Marion	40	29	85	41	859	−4	0	91/74	90/73	88/73	23	77	75	74			97.0	−8.6
Muncie	40	11	85	21	957	−3	2	92/74	90/73	87/73	22	76	76	74				
Peru, Grissom AFB	40	39	86	09	813	−6	−1	90/74	88/73	86/73	22	77	75	74	W 10	SW		
Richmond AP	39	46	84	50	1141	−2	2	92/74	90/74	87/73	22	78	76	75			94.8	−8.5
Shelbyville	39	31	85	47	750	−1	3	93/74	91/74	88/73	22	78	76	75			97.7	−6.0
South Bend AP	41	42	86	19	773	−3	1	91/73	89/73	86/72	22	77	75	74	SW 11	SSW	96.2	−9.2
Terre Haute AP	39	27	87	18	585	−2	4	95/75	92/74	89/73	22	79	77	76	NNW 7	SSW	98.3	−4.9
Valparaiso	41	31	87	02	801	−3	3	93/74	90/74	87/73	22	78	76	75			95.5	−11.0
Vincennes	38	41	87	32	420	1	6	95/75	92/74	90/73	22	79	77	76			100.3	−2.8
IOWA																		
Ames (S)	42	02	93	48	1099	−11	−6	93/75	90/74	87/73	23	78	76	75			97.4	−17.8
Burlington AP	40	47	91	07	692	−7	−3	94/74	91/75	88/73	22	78	77	75	NW 9	SSW	98.6	−11.0
Cedar Rapids AP	41	53	91	42	863	−10	−5	91/76	88/75	86/74	23	78	77	75	NW 9	S	97.7	−15.6
Clinton	41	50	90	13	595	−8	−3	92/75	90/75	87/74	23	78	77	75			97.5	−13.8
Council Bluffs	41	20	95	49	1210	−8	−3	94/75	91/75	88/74	22	78	77	75				
Des Moines AP	41	32	93	39	938	−10	−5	94/75	91/74	88/73	23	78	77	75	NW 11	S	98.2	−14.2
Dubuque	42	24	90	42	1056	−12	−7	90/74	88/73	86/72	22	77	75	74	N 10	SSW	95.2	−15.0
Fort Dodge	42	33	94	11	1162	−12	−7	91/74	88/74	86/72	23	77	75	74	NW 11	S	98.5	−19.1
Iowa City	41	38	91	33	661	−11	−6	92/76	89/76	86/74	22	80	78	76	NW 9	S	97.4	−15.2
Keokuk	40	24	91	24	574	−5	0	95/75	92/75	89/74	22	79	77	76			98.4	−8.8
Marshalltown	42	04	92	56	898	−12	−7	92/76	90/75	88/74	23	78	77	75			98.5	−13.4
Mason City AP	43	09	93	20	1213	−15	−11	90/74	88/74	85/72	24	77	75	74	NW 11		96.5	−21.7
Newton	41	41	93	02	936	−10	−5	94/75	91/74	88/73	23	78	77	75		S	98.2	−14.7
Ottumwa AP	41	06	92	27	840	−8	−4	94/75	91/74	88/73	22	78	77	75			99.1	−12.0
Sioux City AP	42	24	96	23	1095	−11	−7	95/74	92/74	89/73	24	78	77	75	NNW 9	S	99.9	−17.7
Waterloo	42	33	92	24	868	−15	−10	91/76	89/75	86/74	23	78	77	75	NW 9	S	97.7	−19.8

TABLE 1   CLIMATIC CONDITIONS FOR THE UNITED STATES (*Continued*)

State and Station[a]	Lat. ° '	Long. ° '	Elev. Feet	Winter[b] °F Design Dry-Bulb 99%	97.5%	Summer[c] °F Design Dry-Bulb & Mean Coinc. Wet-Bulb 1%	2.5%	5%	Mean Daily Range	Design Wet-Bulb 1%	2.5%	5%	Prevailing Wind Winter (Knots[d])	Summer	Temp. °F Median Annual Extr. Max.	Min.
MICHIGAN																
Adrian	41 55	84 01	754	-1	3	91/73	88/72	85/71	23	76	75	73			97.2	-7.0
Alpena AP	45 04	83 26	610	-11	-6	89/70	85/70	83/69	27	73	72	70	W 5	SW	93.9	-14.8
Battle Creek AP	42 19	85 15	941	1	5	92/74	88/72	85/70	23	76	74	73	SW 8	SW		
Benton Harbor AP	42 08	86 26	643	1	5	91/72	88/72	85/70	20	75	74	72	SSW 8	WSW		
Detroit	42 25	83 01	619	3	6	91/73	88/72	86/71	20	76	74	73	W 11	SW	95.1	-2.6
Escanaba	45 44	87 05	607	-11	-7	87/70	83/69	80/68	17	73	71	69			88.8	-16.1
Flint AP	42 58	83 44	771	-4	1	90/73	87/72	85/71	25	76	74	72	SW 8	SW	95.3	-9.9
Grand Rapids AP	42 53	85 31	784	1	5	91/72	88/72	85/70	24	75	74	72	WNW 8	WSW	95.4	-5.6
Holland	42 42	86 06	678	2		88/72	86/71	83/70	22	75	73	72			94.1	-6.8
Jackson AP	42 16	84 28	1020	1	5	92/74	88/72	85/70	23	76	74	73			96.5	-7.8
Kalamazoo	42 17	85 36	955	1	5	92/74	88/72	85/70	23	76	74	73			95.9	-6.7
Lansing AP	42 47	84 36	873	-3	1	90/73	87/72	84/70	24	75	74	72	SW 12	W	94.6	-11.0
Marquette Co	46 34	87 24	735	-12	-8	84/70	81/69	77/66	18	72	70	68			94.5	-11.8
Mt Pleasant	43 35	84 46	796	0	4	91/73	87/72	84/71	24	76	74	72			95.4	-11.1
Muskegon AP	43 10	86 14	625	2	6	86/72	84/70	82/70	21	75	73	72	E 8	SW		
Pontiac	42 40	83 25	981	0	4	90/73	87/72	85/71	21	76	74	73			95.0	-6.8
Port Huron	42 59	82 25	586	0	4	90/73	87/72	83/71	21	76	74	73	W 8	S		
Saginaw AP	43 32	84 05	667	0	4	91/73	87/72	84/71	23	76	74	72	WSW 7	SW	96.1	-7.6
Sault Ste. Marie AP (S)	46 28	84 22	721	-12	-8	84/70	81/69	77/66	23	72	70	68	E 7	SW	89.8	-21.0
Traverse City AP	44 45	85 35	624	-3	1	89/72	86/71	83/69	22	75	73	71	SSW 9	SW	95.4	-10.7
Ypsilanti	42 14	83 32	716	1	5	92/72	89/71	86/70	22	75	74	72	SW 10	SW		
MINNESOTA																
Albert Lea	43 39	93 21	1220	-17	-12	90/74	87/72	84/71	24	77	75	73			95.1	-28.0
Alexandria AP	45 52	95 23	1430	-22	-16	91/72	88/72	85/70	24	76	74	72			94.5	-36.9
Bemidji AP	47 31	94 56	1389	-31	-26	88/69	85/69	81/67	24	73	71	69	N 8	S		
Brainerd	46 24	94 08	1227	-20	-16	90/73	87/71	84/69	24	75	73	71			90.9	-27.4
Duluth AP	46 50	92 11	1428	-21	-16	85/70	82/69	79/66	22	72	70	68	WNW 12	WSW	95.8	-24.3
Fairbault	44 18	93 16	940	-17	-12	91/74	88/72	85/71	24	77	75	73			96.9	-27.8
Fergus Falls	46 16	96 04	1210	-21	-17	91/72	88/72	85/70	24	76	74	72				
International Falls AP	48 34	93 23	1179	-29	-25	85/68	83/68	80/66	26	71	70	68	N 9	S	93.4	-36.5
Mankato	44 09	93 59	1004	-17	-12	91/72	88/72	85/70	24	77	75	73				
Minneapolis/St. Paul AP	44 53	93 13	834	-16	-12	92/75	89/73	86/71	22	77	75	73	NW 8	S	96.5	-22.0
Rochester AP	43 55	92 30	1297	-17	-12	90/74	87/72	84/71	24	77	75	73	NW 9	SSW		
St. Cloud AP (S)	45 35	94 11	1043	-15	-11	91/74	88/72	85/70	24	76	74	72				
Virginia	47 30	92 33	1435	-25	-21	85/69	83/68	80/66	23	71	70	68			92.6	-33.0
Willmar	45 07	95 05	1128	-15	-11	91/74	88/72	85/71	24	76	74	72			96.8	-24.3
Winona	44 03	91 38	652	-14	-10	91/75	88/73	85/72	24	77	75	74				
MISSISSIPPI																
Biloxi, Keesler AFB	30 25	88 55	26	28	31	94/79	92/79	90/78	16	82	81	80	N 8	S	98	23
Clarksdale	34 12	88 34	178	14	19	96/77	94/77	92/76	21	80	79	78			100.9	13.2
Columbus AFB	33 39	88 27	219	15	20	95/77	93/77	91/76	22	80	79	78	N 7	W	101.6	12.7
Greenville AFB	33 29	90 59	138	15	20	95/77	93/77	91/76	21	80	79	78			99.5	14.9
Greenwood	33 30	90 05	148	15	20	95/77	93/77	91/76	21	80	79	78			100.6	15.3
Hattiesburg	31 16	89 15	148	24	27	96/78	94/77	92/77	21	81	80	79			99.9	18.2
Jackson AP	32 19	90 05	310	21	25	97/76	95/76	93/76	21	79	78	78	NNW 6	NW	99.8	16.0
Laurel	31 40	89 10	236	24	27	96/78	94/77	92/77	21	81	80	79			99.7	17.8
Mccomb AP	31 15	90 28	469	21	26	96/77	94/76	92/76	18	80	79	79				
Meridian AP	32 20	88 45	290	19	23	97/77	95/76	93/76	22	80	79	78	N 6	WSW	98.3	15.7
Natchez	31 33	91 23	195	23	27	96/78	94/78	92/77	21	81	80	79			98.4	18.4
Tupelo	34 16	88 46	361	14	19	96/77	94/77	92/76	22	80	79	78			100.7	11.8
Vicksburg Co	32 24	90 47	262	22	26	97/78	95/78	93/77	21	81	80	79			96.9	18.0
MISSOURI																
Cape Girardeau	37 14	89 35	351	8	13	98/76	95/75	92/75	21	79	78	77			99.5	-6.2
Columbia AP (S)	38 58	92 22	778	-1	4	97/74	94/74	91/73	22	78	77	76	WNW 9	WSW	99.9	-2.1
Farmington AP	37 46	90 24	928	3	8	96/76	93/75	90/74	22	78	77	75			98.4	-7.6
Hannibal	39 42	91 21	489	-2	3	96/76	93/76	90/76	22	80	78	77	NNW 11	SSW	101.2	-6.1
Jefferson City	38 34	92 11	640	2	7	98/75	95/74	92/74	23	78	77	76				
Joplin AP	37 09	94 30	980	6	10	100/73	97/73	94/73	24	78	77	76	NNW 12	SSW		
Kansas City AP	39 07	94 35	791	2	6	99/75	96/74	93/74	20	78	77	76	NW 9	S	100.2	-4.3
Kirksville AP	40 06	92 33	964	-5	0	96/74	93/74	90/73	24	78	77	76			98.3	-10.8
Mexico	39 11	91 54	775	-1	4	97/74	94/74	91/73	22	78	77	76			101.2	-8.0
Moberly	39 24	92 26	850	-2	3	97/74	94/74	91/73	23	78	77	76				
Poplar Bluff	36 46	90 25	380	11	16	98/78	95/76	92/76	22	81	79	78				
Rolla	37 59	91 43	1204	3	8	94/77	91/75	89/74	22	78	77	76			99.4	-3.1
St. Joseph AP	39 46	94 55	825	-3	2	96/77	93/76	91/76	23	81	79	77	NNW 9	S	100.6	-8.0
St. Louis AP	38 45	90 23	535	2	6	97/75	94/75	91/74	21	78	77	76	NW 9	WSW		
St. Louis Co	38 39	90 38	462	3	8	98/75	94/75	91/74	18	78	77	76	NW 6		99.1	-2.7
Sikeston	36 53	89 36	325	9	15	98/77	95/76	92/75	21	80	78	77				
Sedalia, Whiteman AFB	38 43	93 33	869	-1	4	95/76	92/75	90/75	22	79	78	76	NNW 7	SSW	100.0	-5.1
Sikeston	36 53	89 36	325	9	15	98/77	95/76	92/75	21	80	78	77				
Springfield AP	37 14	93 23	1268	3	9	96/73	93/74	91/74	23	78	77	75	NNW 10	S	97.2	-2.4

Equipment Operation

TABLE 1 CLIMATIC CONDITIONS FOR THE UNITED STATES (Continued)

Col. 1	Col. 2		Col. 3		Col. 4	Col. 5 Winter,[b] °F		Col. 6 Summer,[c] °F			Col. 7	Col. 8			Col. 9 Prevailing Wind		Col. 10 Temp. °F	
State and Station[a]	Lat.		Long.		Elev.	Design Dry-Bulb		Design Dry-Bulb and Mean Coincident Wet-Bulb			Mean Daily	Design Wet-Bulb					Median of Annual Extr.	
	°	'	°	'	Feet	99%	97.5%	1%	2.5%	5%	Range	1%	2.5%	5%	Winter	Summer	Max.	Min.
															Knots[d]			
Roswell, Walker AFB	33	18	104	32	3676	13	18	100/66	98/66	96/66	33	71	70	69	N 6	SSE	103.0	2.7
Santa Fe Co	35	37	106	05	6307	6	10	90/61	88/61	86/61	28	63	62	61			90.1	-1.2
Silver City AP	32	38	108	10	5442	5	10	95/61	94/60	91/60	30	66	64	63				
Socorro AP	34	03	106	53	4624	13	17	97/62	95/62	93/62	30	67	66	65				
Tucumcari AP	35	11	103	36	4039	8	13	99/66	97/66	95/65	28	70	69	68	NE 8	SW	102.7	1.1
NEW YORK																		
Albany AP (S)	42	45	73	48	275	-6	-1	91/73	88/72	85/70	23	75	74	72	WNW 8	S		
Albany Co	42	39	73	45	19	-4	1	91/73	88/72	85/70	20	75	74	72			95.2	-11.4
Auburn	42	54	76	32	715	-3	2	90/73	87/71	84/70	22	75	73	72			92.4	-9.5
Batavia	43	00	78	11	922	1	5	90/72	87/71	84/70	22	75	73	72			92.2	-7.5
Binghamton AP	42	13	75	59	1590	-2	1	86/71	83/69	81/68	20	73	72	70	WSW 10	WSW	92.9	-9.3
Buffalo AP	42	56	78	44	705	2	6	88/71	85/70	83/69	21	74	73	72	W 10	SW	90.0	-3.2
Cortland	42	36	76	11	1129	-5	0	88/71	85/71	82/70	23	74	73	71			93.8	-11.2
Dunkirk	42	29	79	16	692	4	9	88/73	85/72	83/71	18	75	74	72	SSW 10	WSW		
Elmira AP	42	10	76	54	955	-4	1	89/71	86/71	83/70	24	74	73	71			96.2	-6.7
Geneva (S)	42	45	76	54	613	-3	2	90/73	87/71	84/70	22	75	73	72			96.1	-6.5
Glens Falls	43	20	73	37	328	-11	-5	88/72	85/71	82/69	23	74	73	71	NNW 6	S		
Gloversville	43	02	74	21	760	-8	-2	89/72	86/71	83/69	23	75	74	72			93.2	-14.6
Hornell	42	21	77	42	1325	-4	0	88/71	85/70	82/69	24	74	73	72				
Ithaca (S)	42	27	76	29	928	-5	0	88/71	85/71	82/70	24	74	73	71	W 6	SW		
Jamestown	42	07	79	14	1390	-1	3	88/70	86/70	83/69	20	74	72	71	WSW 9	WSW		
Kingston	41	56	74	00	279	-3	2	91/73	88/72	85/70	22	76	74	73				
Lockport	43	09	79	15	638	4	7	89/74	86/72	84/71	21	76	74	73	N 9	SW	92.2	-4.8
Massena AP	44	56	74	51	207	-13	-8	86/70	83/69	80/68	20	73	72	70				
Newburgh, Stewart AFB	41	30	74	06	471	-1	4	90/73	88/72	85/70	21	76	74	73	W 10	W		
NYC-Central Park (S)	40	47	73	58	157	11	15	92/74	89/73	87/72	17	76	75	74			94.9	3.8
NYC-Kennedy AP	40	39	3	47	13	12	15	90/73	87/72	84/71	16	76	75	74	WNW 4	SSW		
NYC-La Guardia AP	40	46	73	54	11	11	15	92/74	89/73	87/72	16	76	75	74	WNW 15	SW		
Niagara Falls AP	43	06	79	57	590	4	7	89/74	86/72	84/71	20	76	74	73	W 9	SW		
Olean	42	14	78	22	2119	-2	2	87/71	84/71	81/70	23	74	73	71				
Oneonta	42	31	75	04	1775	-7	-4	86/71	83/69	80/68	24	73	72	70				
Oswego Co	43	28	76	33	300	1	7	86/73	83/71	80/70	20	75	73	72	E 7	WSW	91.3	-7.4
Plattsburg AFB	44	39	73	28	235	-13	-8	86/70	83/69	80/68	22	73	72	70	NW 6	SE		
Poughkeepsie	41	38	73	55	165	0	6	92/74	89/74	86/72	21	77	75	74	NNE 6	SSW	98.1	-5.6
Rochester AP	43	07	77	40	547	1	5	91/73	88/71	85/70	22	75	73	72	WSW 11	WSW		
Rome, Griffiss AFB	43	14	75	25	514	-11	-5	88/71	85/70	83/69	22	75	73	71	NW 5	W		
Schenectady (S)	42	51	73	57	377	-4	1	90/73	87/72	84/70	22	75	74	72	WNW 8	S		
Suffolk County AFB	40	51	72	38	67	7	10	86/72	83/71	80/70	16	76	74	73	NW 9	SW		
Syracuse AP	43	07	76	07	410	-3	2	90/73	87/71	84/70	20	75	73	72	N 7	WNW	93.	-10.0
Utica	43	09	75	23	714	-12	-6	88/73	85/71	82/70	22	75	73	71	NW 12	W		
Watertown	43	59	76	01	325	-11	-6	86/73	83/71	81/70	20	75	73	72	E 7	WSW	91.7	-19.6
NORTH CAROLINA																		
Asheville AP	35	26	82	32	2140	10	14	89/73	87/72	85/71	21	75	74	72	NNW 12	NNW	91.9	5.8
Charlotte AP	35	13	80	56	736	18	22	95/74	93/74	91/74	20	77	76	76	NNW 6	SW	97.8	12.6
Durham	35	52	78	47	434	16	20	94/75	92/75	90/75	20	78	77	76			98.9	9.6
Elizabeth City AP	36	16	76	11	12	12	19	93/78	91/77	89/76	18	80	78	78	NW 8	SW		
Fayetteville, Pope AFB	35	10	79	01	218	17	20	95/76	92/76	90/75	20	79	78	77	N 6	SSW	99.1	13.1
Goldsboro, Seymour-Johnson	35	20	77	58	109	18	21	94/77	91/76	89/75	18	79	78	77	N 8	SW	99.8	13.0
Greensboro AP (S)	36	05	79	57	897	14	18	93/74	91/73	89/73	21	77	76	75	NE 7	SW	97.7	9.7
Greenville	35	37	77	25	75	18	21	93/77	91/76	89/75	19	79	78	77				
Henderson	36	22	78	25	480	12	15	95/77	92/76	90/76	20	79	78	77				
Hickory	35	45	81	23	1187	14	18	92/73	90/72	88/72	21	75	74	73			96.5	9.6
Jacksonville	34	50	77	37	95	20	24	92/78	90/78	88/77	18	80	79	78				
Lumberton	34	37	79	04	129	18	21	95/76	92/76	90/75	20	79	78	77				
New Bern AP	35	05	77	03	20	20	24	92/78	90/78	88/77	18	80	79	78			98.2	15.1
Raleigh/Durham AP (S)	35	52	78	47	434	16	20	94/75	92/75	90/75	20	78	77	76	N 7	SW	97.7	12.2
Rocky Mount	35	58	77	48	121	18	21	94/77	91/76	89/75	19	79	78	77				
Wilmington AP	34	16	77	55	28	23	26	93/79	91/78	89/77	18	81	80	79	N 8	SW	96.9	18.2
Winston-Salem AP	36	08	80	13	969	16	20	94/74	91/73	89/73	20	76	75	74	NW 8	WSW		
NORTH DAKOTA																		
Bismarck AP (S)	46	46	100	45	1647	-23	-19	95/68	91/68	88/67	27	73	71	70	WNW 7	S	100.3	-31.5
Devils Lake	48	07	98	54	1450	-25	-21	91/69	88/68	85/66	25	73	71	69			97.5	-30.4
Dickinson AP	46	48	102	48	2585	-21	-17	94/68	90/66	87/65	25	71	69	68	WNW 12	SSE	101.0	-31.3
Fargo AP	46	54	96	48	896	-22	-18	92/73	89/71	85/69	25	76	74	72	SSE 11	S	97.3	-29.7
Grand Forks AP	47	57	97	24	911	-26	-22	91/70	87/70	84/68	25	74	72	70	N 8	S	97.6	-29.0
Jamestown AP	46	55	98	41	1492	-22	-18	94/70	90/69	87/68	26	74	74	71			101.3	-27.9
Minot AP	48	25	101	21	1668	-24	-20	92/68	89/67	86/65	25	72	70	68	WSW 10	S		
Williston	48	09	103	35	1876	-25	-21	91/68	88/67	85/65	25	72	70	68			99.7	-32.9

Control Theory   Chap. 3

TABLE 1 CLIMATIC CONDITIONS FOR THE UNITED STATES (*Concluded*)

Col. 1	Col. 2 Lat.		Col. 3 Long.		Col. 4 Elev.	Col. 5 Winter,[b] °F Design Dry-Bulb		Col. 6 Summer,[c] °F Design Dry-Bulb and Mean Coincident Wet-Bulb			Col. 7 Mean Daily	Col. 8 Design Wet-Bulb			Col. 9 Prevailing Wind			Col. 10 Temp. °F Median of Annual Extr.	
State and Station[a]	°	′	°	′	Feet	99%	97.5%	1%	2.5%	5%	Range	1%	2.5%	5%	Winter (Knots[d])	Summer		Max.	Min.
VIRGINIA																			
Charlottesville	38	02	78	31	870	14	18	94/74	91/74	88/73	23	77	76	75	NE 7	SW		97.4	8.0
Danville AP	36	34	79	20	590	14	16	94/74	92/73	90/73	21	77	76	75				100.1	9.2
Fredericksburg	38	18	77	28	100	10	14	96/76	93/75	90/74	21	78	77	76					
Harrisonburg	38	27	78	54	1370	12	16	93/72	91/72	88/71	23	75	74	73					
Lynchburg AP	37	20	79	12	916	12	16	93/74	90/74	88/73	21	77	76	75	NE 7	SW		97.2	7.6
Norfolk AP	36	54	76	12	22	20	22	93/77	91/76	89/76	18	79	78	77	NW 10	SW		97.2	15.3
Petersburg	37	11	77	31	194	14	17	95/76	92/76	90/75	20	79	78	77					
Richmond AP	37	30	77	20	164	14	17	95/76	92/76	90/75	21	79	78	77	N 6	SW		97.9	9.6
Roanoke AP	37	19	79	58	1193	12	16	93/72	91/72	88/71	23	75	74	73	NW 9	SW			
Staunton	38	16	78	54	1201	12	16	93/72	91/72	88/71	23	75	74	73	NW 9	SW		95.9	2.5
Winchester	39	12	78	10	760	6	10	93/75	90/74	88/74	21	77	76	75				97.3	3.7
WASHINGTON																			
Aberdeen	46	59	123	49	12	25	28	80/65	77/62	73/61	16	65	63	62	ESE 6	NNW		91.9	19.3
Bellingham AP	48	48	122	32	158	10	15	81/67	77/65	74/63	19	68	65	63	NNE 15	WSW		87.4	10.3
Bremerton	47	34	122	40	162	21	25	82/65	78/64	75/62	20	66	64	63	E 8	N			
Ellensburg AP	47	02	120	31	1735	2	6	94/65	91/64	87/62	34	66	65	63					
Everett, Paine AFB	47	55	122	17	596	21	25	80/65	76/64	73/62	20	67	64	63	ESE 6	NNW		84.9	15.2
Kennewick	46	13	119	08	392	5	11	99/68	96/67	92/66	30	70	68	67				103.4	2.0
Longview	46	10	122	56	12	19	24	88/68	85/67	81/65	30	69	67	66	ESE 9	NW		96.0	14.8
Moses Lake, Larson AFB	47	12	119	19	1185	1	7	97/66	94/65	90/63	32	67	66	64	N 8	SSW			
Olympia AP	46	58	122	54	215	16	22	87/66	83/65	79/64	32	67	66	64	NE 4	NE		83.5	19.4
Port Angeles	48	07	123	26	99	24	27	72/62	69/61	67/60	18	64	62	61					
Seattle-Boeing Field	47	32	122	18	23	21	26	84/68	81/66	77/65	24	69	67	65				90.2	22.0
Seattle Co (S)	47	39	122	18	20	22	27	85/68	82/66	78/65	19	69	67	65	N 7	N		90.1	19.9
Seattle-Tacoma AP (S)	47	27	122	18	400	21	26	84/65	80/64	76/62	22	66	64	63	E 9	N		98.8	-4.9
Spokane AP (S)	47	38	117	31	2357	-6	2	93/64	90/63	87/62	28	65	64	62	NE 6	SW		89.4	18.8
Tacoma,McChord AFB	47	15	122	30	100	19	24	86/66	82/65	79/63	22	68	66	64	S 5	NNE		103.0	3.8
Walla Walla AP	46	06	118	17	1206	0	7	97/67	94/66	90/65	27	69	67	66	W 5	W		101.1	1.0
Wenatchee	47	25	120	19	632	7	11	96/67	96/66	92/64	32	68	67	65					
Yakima AP	46	34	120	32	1052	-2	5	96/65	93/65	89/63	36	68	66	65	W 5	NW			
WEST VIRGINIA																			
Beckley	37	47	81	07	2504	-2	4	83/71	81/69	79/69	22	73	71	70	WNW 9	WNW			
Bluefield AP	37	18	81	13	2867	-2	4	83/71	81/69	79/69	22	73	71	70					
Charleston AP	38	22	81	36	939	7	11	92/74	90/73	87/72	20	76	75	74	SW 8	SW		97.2	2.9
Clarksburg	39	16	80	21	977	6	10	92/74	90/73	87/72	21	76	75	74					
Elkins AP	38	53	79	51	1948	1	6	86/72	84/70	82/70	22	74	72	71	WNW 9	WNW		90.6	-7.3
Huntington Co	38	25	82	30	565	5	10	94/76	91/74	89/73	22	78	77	75	W 6	W		97.1	2.1
Martinsburg AP	39	24	77	59	556	6	10	93/75	90/74	88/74	21	77	76	75	WNW 10	W		99.0	1.1
Morgantown AP	39	39	79	55	1240	4	8	90/74	87/73	85/73	22	76	75	74					
Parkersburg Co	39	16	81	34	615	7	11	93/75	90/74	88/73	21	77	76	75	WSW 7	WSW		95.9	.7
Wheeling	40	07	80	42	665	1	5	89/72	86/71	84/70	21	74	73	72	WSW 10	WSW		97.5	-.6
WISCONSIN																			
Appleton	44	15	88	23	730	-14	-9	89/74	86/72	83/71	23	76	74	72				94.6	-16.2
Ashland	46	34	90	58	650	-21	-16	85/70	82/68	79/66	23	72	70	68				94.1	-26.8
Beloit	42	30	89	02	780	-7	-3	92/75	90/75	88/74	24	78	77	75					
Eau Claire AP	44	52	91	29	888	-15	-11	92/75	89/73	86/71	23	77	75	73					
Fond Du Lac	43	48	88	27	760	-12	-8	89/74	86/72	84/71	23	76	74	72				96.0	-17.7
Green Bay AP	44	29	88	08	682	-13	-9	88/74	85/72	83/71	23	76	74	72	W 8	SW		94.3	-17.9
La Crosse AP	43	52	91	15	651	-13	-9	91/75	88/73	85/72	22	77	75	74	NW 10	S		95.7	-21.3
Madison AP (S)	43	08	89	20	858	-11	-7	91/74	88/73	85/71	22	77	75	73	NW 8	SW		93.6	-16.8
Manitowoc	44	06	87	41	660	-11	-7	89/74	86/72	83/71	21	76	74	72				94.1	-13.7
Marinette	45	06	87	38	605	-15	-11	87/73	84/71	82/70	20	75	73	71				95.9	-15.8
Milwaukee AP	42	57	87	54	672	-8	-4	90/74	87/73	84/71	21	76	74	73	WNW 10	SSW			
Racine	42	43	87	51	730	-6	-2	91/75	88/73	85/72	21	77	75	74					
Sheboygan	43	45	87	43	648	-10	-6	89/75	86/73	83/72	20	77	75	74				97.0	-12.4
Stevens Point	44	30	89	34	1079	-15	-11	92/75	89/73	86/71	23	77	75	73				95.3	-24.1
Waukesha	43	01	88	14	860	-9	-5	90/74	87/73	84/71	22	76	74	73				95.7	-14.3
Wausau AP	44	55	89	37	1196	-16	-12	91/74	88/72	85/70	23	76	74	72					
WYOMING																			
Casper AP	42	55	106	28	5338	-11	-5	92/58	90/57	87/57	31	63	61	60	NE 10	SW		97.3	-20.9
Cheyenne	41	09	104	49	6126	-9	-1	89/58	86/58	84/57	30	63	62	60	N 11	WNW		92.5	-15.9
Cody AP	44	33	109	04	4990	-19	-13	89/60	86/60	83/59	32	64	63	61				97.4	-21.9
Evanston	41	16	110	57	6780	-9	-3	86/55	84/55	82/54	32	59	58	57				89.2	-21.2
Lander AP (S)	42	49	108	44	5563	-16	-11	91/61	88/61	85/60	32	64	63	61	E 5	NW		94.9	-22.6
Laramie AP (S)	41	19	105	41	7266	-14	-6	84/56	81/56	79/55	28	61	60	59					
Newcastle	43	51	104	13	4265	-17	-12	91/64	87/63	84/63	30	69	68	66				99.4	-19.0
Rawlins	41	48	107	12	6740	-12	-4	86/57	83/57	81/56	40	62	61	60					
Rock Springs AP	41	36	109	04	6745	-9	-3	86/55	84/55	82/54	32	59	58	57	WSW 10	W			
Sheridan AP	44	46	106	58	3964	-14	-8	94/62	91/62	88/61	32	66	65	63	NW 7	N		99.8	-23.6
Torrington	42	05	104	13	4098	-14	-8	94/62	91/62	88/61	30	66	65	63				101.1	-20.7

Equipment Operation

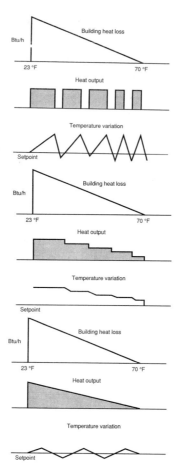

Figure 3-3 Each mode of control system operation gives a particular level of control sensitivity.

The control of airflow in a building is basically a function of blowers used for circulating the air and, in some cases, dampers for control of outside air used as ventilation. Blowers are either run continuously or are turned on and off when the heating or air-conditioning system is turned on or off. Damper control is usually either by on–off controls or by modulating controls.

CONTROL SYSTEM OPERATION

As control systems are matched with heating and cooling units to control the temperature in a building, certain types of control systems lend themselves naturally to applications with certain types of heating and cooling equipment.

Digital Control

Digital control operates heating and air-conditioning equipment, so it operates in on–off cycles. When a heating or air-conditioning unit is on, it produces heat or cooling at a maximum rate, and when it is off, it produces no effect. At any outdoor temperature other than the maximum for a given area, the equipment must be cycled on and off to produce the wanted amount of heat or cooling over a period of time. By turning the equipment on and off in fairly short cycles, the heating or cooling output for 1 hour can be made to match the load during that hour (Figure 3-4).

Example The control device used to cycle a heating unit on and off in an electrical control system is called a *thermostat.* A thermostat is a switching device with a sensor that senses ambient temperature. *Ambient temperature* is the temperature of the air surrounding the device. The sensor opens or closes an electrical switch that is wired in series with the heating unit control circuit (Figure 3-5). The thermostat has a set-point adjustment on which the desired space temperature is set. If the ambient temperature varies from the set-point temperature by more than about 2 degrees plus or minus, the electrical switch is activated. If the temperature goes above the set-point temperature by 2 degrees, the switch opens, and if it goes below the set-point temperature by 2 degrees, the switch closes. When the switch opens, the heating unit is turned off, and when the temperature goes below the set point, the heating unit is turned on. With the switch wired into the heating unit control circuit, the thermostat controls operation of the heating unit.

When outside air is used as ventilation air for a building it is usually controlled by dampers in an air duct from the outside to the return-air side of the air distribution system for the building (Figure 3-6). When the outside-air damper is controlled by an on–off control system, the damper is operated by a two-position damper motor. The motor is normally controlled so that it opens the damper when the system circulation blower is running and closes it when

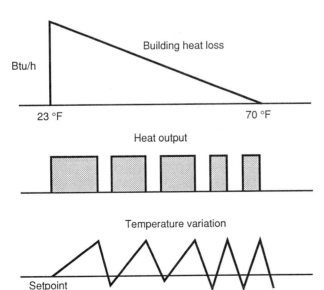

Figure 3-4 Digital control provides on–off control cycles to provide the average amount of heating or cooling needed in 1 hour.

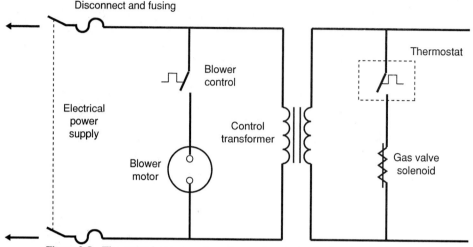

Figure 3-5   The control components and operating parts of a heating or air-conditioning system are shown on a schematic wiring diagram of the unit.

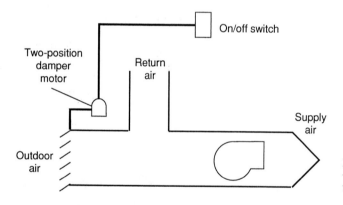

Figure 3-6   Digital damper control provides on–off control of the damper blades.

the blower is not running. Conditions in a controlled space maintained by digital or on–off controls will fluctuate around the set-point condition. Conditions in the space can be kept within acceptable limits for comfort by using a low-voltage system with very sensitive control elements.

Stepped Control

Stepped control is the control of a heating or air-conditioning system in such a way that the heat or cooling output comes on in stages or increments of the total output potential. Stepped control is especially useful in controlling a unit such as an electric heating unit. Most electric heating units have several heating elements. Each element provides a part of the total output of the unit. To operate such a unit with a stepped control system, each element, or some combination of elements, is operated by one stage of the controller. The stages are brought on in sequence with either a multistage thermostat or a sequencer.

A *multistage thermostat* has more than one sensor and operating switch (Figure 3-7). Each switch is controlled by one of the sensors. There is a built-in differential between the temperatures at which the various sensors operate the switches. Each switch operates one of the stages in the unit being controlled. A *sequencer* is a control device in which electrical switches are opened or closed in sequence, with a time interval between opening and closing (Figure 3-8).

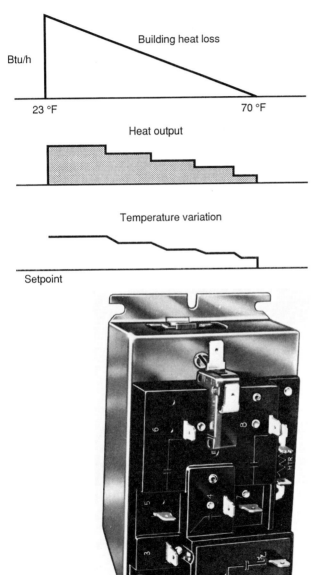

Figure 3-7 Multistage control makes it possible to activate heating or air-conditioning equipment on in steps to provide the control function required.

Figure 3-8 An electrical sequencer is one device used to achieve step control.

Two types of sequencers are commonly used: a time-delay relay with several contacts, and a mechanical sequencer.

In a *time-delay relay sequencer* the contacts make and break in sequence, with a time interval of from 30 to 60 seconds between stages. Each stage of the device being controlled is wired in series through one of the sets of contacts in the sequencer. When the sequencer is energized by a controller the stages are turned on one at a time. They are staged off in the reverse sequence when the controller is satisifed.

A *mechanical sequencer* is a device in which a small motor turns a shaft on which cams turn switches on and off in sequence (Figure 3-9). The switches are arranged so that they turn on and off with whatever time interval is desired in relation to the rotational speed of the motor. A sequencer is activated when a thermostat or other control calls for action. The switches are connected in series in line or control circuits to the different elements in the equipment being controlled. The sequencer is started and stopped by a controller that senses space conditions.

> **Example** A two-stage air-conditioning unit has either a two-speed compressor motor or two compressors. To operate a unit with a two-speed compressor motor a stepped control system brings on each speed of the motor by one stage of the controller. If a unit is used with two compressors, each compressor is brought on by one stage of the controller. To achieve this control, a two-stage thermostat is used. The thermostat has two sensors, each controlling a switch. A differential is built in to the thermostat between the two switches. The first stage "makes" at about a 2-degree difference between the set point and the ambient temperature, and the second stage makes at about a 4-degree difference. If the ambient temperature goes 2 degrees above the set-point temperature the first stage makes,

Figure 3-9 A mechanical sequencer achieves step control by the opening or closing of electrical switches in sequence.

Control Theory   Chap. 3

and if the temperature goes 4 degrees above the set-point temperature, the second stage makes.

A stepped control system provides control of the conditions in building spaces within tolerances acceptable to most people. Equipment controlled by a stepped system usually operates more efficiently than equipiment in an on–off system. The equipmient output is more closely matched to the building loads.

Proportional Control

Proportional control is a control system that controls a heating or air-conditioning system in such a way that at any given time, heating or cooling output just matches the load on the spaces being conditioned (Figure 3-10). In a proportional control system the output signal from an actuator is proportional to the difference between the set-point condition and the space condition.

In most electric and electronic control systems a proportional output signal is achieved by the use of a special wiring circuit called a *bridge circuit* (Figure 3-11). A pneumatic system naturally controls in a proportional manner. Electric or electronic control systems utilize a bridge circuit in the controller which produces an output signal that is proportional to any deviation from set point in the controlled variable. Input is from a sensor in the controller that senses changes in the controlled variable and transmits these changes through a variable resistor to the actuator in the system. The changes are transmitted to a second resistor, located in the actuator. The actuator controls the heating or cooling unit to bring the controlled variable back to a desired condition.

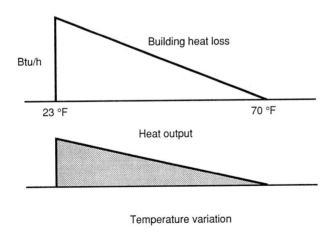

Figure 3-10 Proportional control provides just the amount of heating or cooling effect to match the load at any given time.

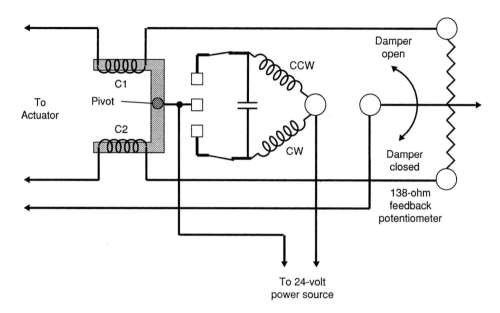

Figure 3-11 A bridge circuit is typically part of an electrical proportional control system, with the sensor controller one half of the circuit and the controller the other half.

Example When a proportional controller is used in an electric control system, the sensor in the thermostat (controller) moves a wiper back and forth across a resistance coil. The wiper blade is connected to one power lead for the control system. As the blade moves it causes the resistance coil to divide the amperage in divided circuits. The two ends of the coil are connected by control wiring to a balancing relay in the system operator. A balancing relay is a two-pole relay with a contact that can move back and forth from one pole to the other. The movement of the contact between the two poles is controlled by amperage from the controller. The balancing relay operates a bidirectional damper or valve motor so that it runs in one direction or the other (Figure 3-12). As the motor moves from one position to another, it moves a wiper blade across a resistance coil similar to the one in the controller. This wiper blade is connected to the second lead, from the power supply to the control system. As the wiper blade moves across the resistor on the motor it will reach a position in which the output from the controller is just balanced. Since the motor also drives an actuator of some sort, this actuator controls the heating or cooling unit to produce the controlling effect desired.

Proportional control is an inherent part of a pneumatic control system. In a pneumatic control system control air pressure varies in proportion to the difference between the set-point condition and space conditions. The operators used in a pneumatic control system provide proportional operation in the way they function (Figure 3-13).

Proportional control of outside-air dampers is often used to control the

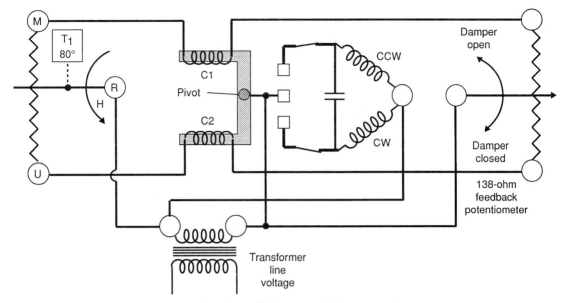

Figure 3-12 Proportional control allows control of dampers, valves, or sequencers.

amount of ventilation air introduced to a building. The control may be one of two types. It may be control of a damper on the outside-air duct only, or it may be control of both outside and return air. Control of outside air only is achieved by using a damper in the outside-air duct that is controlled by a modulating damper motor. In an electromechanical control system the damper motor is controlled by a remote potentiometer. The potentiometer is an adjustable switch that can be set manually to control the damper opening. In a pneumatic control system the adjustable switch is a manual control that modulates the control air pressure going to the damper operator (Figure 3-14).

Modulating control of outside air for ventilation is usually used in building spaces in which the people load varies considerably, such spaces as auditoriums or in rooms whose use by large numbers of people is cyclical. Control of outside air and return air, to provide a given mixed-air temperature, is achieved by the use of a modulating damper motor that controls both dampers at the same time (Figure 3-15). The dampers are operated so that their blades are always 90° opposed to each other. When one damper opens, the other closes. The damper motor is controlled by a mixed-air controller mounted in the mixed-air duct downstream from the two dampers.

Control of both outside air and return air is generally used in buildings in which a cooling load occurs at the same time as cool outside temperatures occur. This usually means a building with rather high people loads and associated loads such as lights and appliances. In this control scheme outside air is used to cool the main building when the outside-air temperature is cool enough to be used for that purpose. Conditions in building spaces can be maintained

Control System Operation

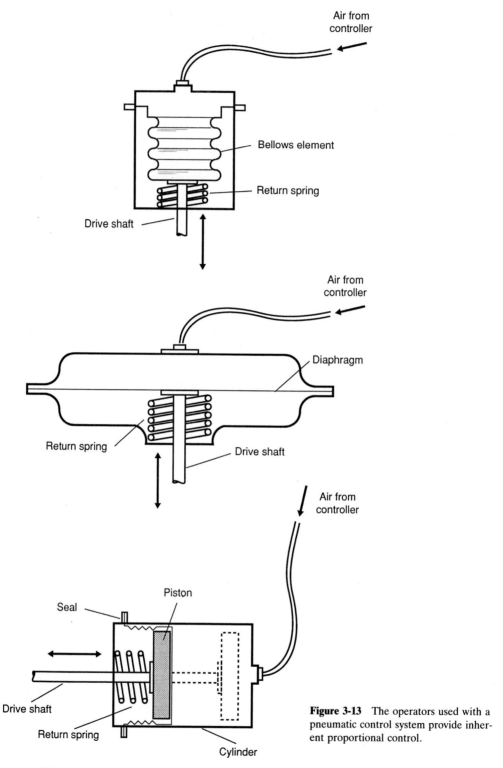

Air from
controller

Bellows element

Return spring

Drive shaft

Air from
controller

Diaphragm

Return spring

Drive shaft

Air from
controller

Piston

Seal

Drive shaft

Return spring

Cylinder

Figure 3-13 The operators used with a
pneumatic control system provide inherent proportional control.

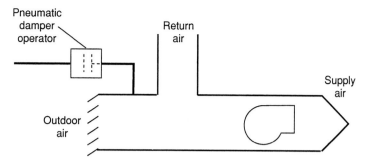

Figure 3-14 A damper control system using pneumatic controls functions basically as a proportional control system.

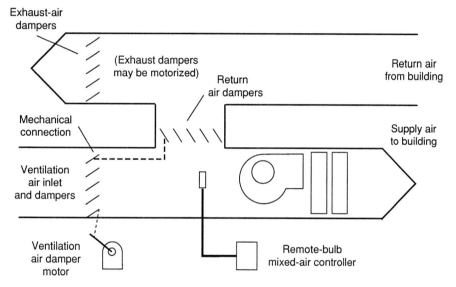

Figure 3-15 An economizer damper control system functions well with a pneumatic control system.

closer to set-point conditions with proportional control than with either of the other types of control systems.

SUMMARY

Controls and systems of control are used to provide the proper output of heat, cooling effect, humidity, or air control to maintain comfortable conditions in a building. The equipment that provides the heating, air conditioning, humidity, or air control must be sized properly, and the controls must be chosen to match the equipment to provide the control desired.

In matching control systems to equipment a technician must know how various types of equipment operate. Some equipment is simply turned on or off in cycles. In a digital system the cycles vary so that the proper amount of control effect occurs over a given time. Some equipment operates in steps, each step providing some increment of control to meet a load. This is called stepped control. One other type of system modulates equipment output so that it just matches the load at any given time. This is called proportional control. Each of the foregoing three types of equipment operation is matched by control systems that provide the type of control that matches the equipment output to a load over a given length of time.

QUESTIONS

3-1. Name three ways in which equipment is controlled.

3-2. What type of control system is most often used with small residential heating and air-conditioning systems?

3-3. How does a stepped controller work to maintain a given temperature in a building?

3-4. What type of control system works best with hydronic heating or air-conditioning equipment?

3-5. *True or false:* A thermostat used in an electric control system could be called a temperature-actuated electrical switch.

3-6. Match the term in the first column with the phrase that best matches it in the second column by placing the letter that precedes the description in the space provided preceding the term.

    A. _____ Digital   a. controls in steps
    B. _____ Stepped   b. always matches load
    C. _____ Proportional   c. on and off

3-7. An air-conditioning system with two compressors can be operated as a _____ (stepped/proportional) system.

3-8. *True or false:* The sensor in a bridge circuit arrangement to provide proportional control for electric or electronic control systems moves a wiper blade across a resistor.

3-9. *True or false:* A wiper blade thatis driven across a resistor by the actuator motor in a bridge circuit provides a balancing effect in the control circuit.

3-10. *True or false:* Two-position control of an outside air damper is achieved by use of a modulating control system.

3-11. *True or false:* Conditions in a building can be maintained closer to set-point conditions by use of proportional control than with the other types of control systems.

APPLICATION EXERCISES

3-1. Match the term in the first column with the phrase that best matches it in the second column by placing the letter that precedes the description in the space provided preceding the term.

A. _____ Digital   a. electric furnace
B. _____ Stepped   b. hydronic system
C. _____ Proportional   c. gas or oil furnace
                                    d. air-conditioning unit

3-2. Draw in the logical wiring sequence on the attached diagram of a thermostat used in an electrical control system operating a gas furnace.

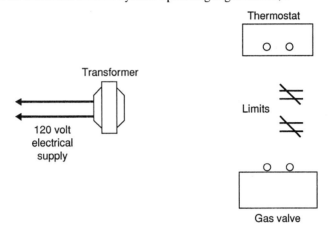

3-3. Draw in the logical wiring sequence on the attached diagram of a thermostat controlled sequencer used with a step controller in an electrical control system operating an electric furnace.

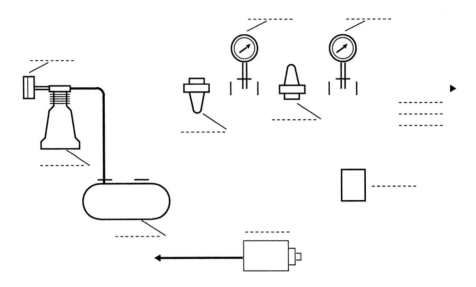

3-4. Draw in the logical air piping for control on the attached diagram of a pneumatic control system used on a hydronic heating system.

4

System Elements

All control systems, regardless of type, are composed of certain operative elements. The elements differ somewhat depending on the type of system, and we give them different names for the different systems, but they each exist in some form in each system. Each element of a control system provides certain functions within the system that are necessary and important for system operation. Sophisticated systems have some additional components, such as central control panels and cumulators, but these are used in particular systems for special purposes. The main elements of a control system are the controller, a signal, an actuator, and feedback (Figure 4-1).

CONTROLLERS

Controllers are devices in a control system that sense conditions and originate signals to convey information about those conditions. Generally, a controller is designed to sense differences between set-point conditions and controlled variable conditions, and to send a signal relative to those differences. Controllers are usually called by different names in the various types of control systems. In electromechanical systems controllers are called *thermostats, humidistats,* or *pressurestats,* depending on the variable controlled. The same device

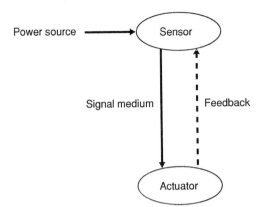

Figure 4-1 A typical control system contains four elements: a controller, a signal, an actuator, and feedback.

in a pneumatic system may be called a *relay, thermostat, humidistat, controller,* or *transmitter*. In an electronic control system the same device may be called a *sensor, transmitter,* or *controller*.

Electromechanical Controllers

Electromechanical controllers are used in a control system using alternating-current (ac) electricity as a control signal. Most electromechanical controllers utilize some form of electrical switch as a control mechanism. Electromechanical controls are often used in controlling equipment in residential or small commercial applications.

Electromechanical temperature control. Electromechanical controls are often used in temperature control systems. The controller used in an electromechanical temperature control system is called a *thermostat*. Basically, two types of thermostats are used: room thermostats and remote thermostats. A *room thermostat* is located in the space in which temperature control is desired and is connected to a heating or cooling system by electrical control wiring. A typical room thermostat contains a sensor that is affected by temperature changes. In response to changes in the temperature the sensor actuates an electrical switch. A typical sensor is made of two strips of metal with different coefficients of expansion, fused together. This is called a *bimetallic element* (Figure 4-2). With a temperature drop the strips of metal warp one way, and with a temperature rise they warp the other way. The warping action activates an electrical switch through a mechanical linkage.

A thermostat has a set-point indicator that is set to the temperature desired in the space in which the thermostat is located. Any deviation of temperature in the space from the set-point temperature causes the sensor to actuate an electrical switch. In an electromechanical system this switch is wired in series in an electrical circuit that runs from the control system electrical power source, usually a transformer, to an actuator that controls the heat source.

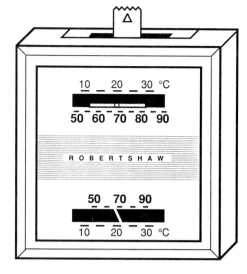

Figure 4-2 A typical controller is a thermostat that contains a bimetallic sensor.

The second type of thermostat is a *remote-bulb thermostat.* This type of thermostat has a sensing element that is remote from the switching mechanism (Figure 4-3). The sensor in a remote-bulb thermostat is a metal bulb filled with a volatile fluid such as a refrigerant. The bulb is connected to the switching

Figure 4-3 A thermostat can be used that senses conditions some distance away from the control device itself.

device by a capillary tube. The temperature around the bulb controls the pressure of the fluid within the bulb. If the temperature around the bulb rises, some of the fluid in the bulb evaporates and the pressure in the capillary goes up. If the temperature around the bulb goes down, some of the fluid condenses and the pressure in the capillary is decreased. The capillary tube running from the sensing bulb is connected to a diaphragm or bellows in the switching device. A connecting linkage runs from the bellows or diaphragm to a lever arrangement that opens or closes an electrical switch. The switch is wired in series in a control circuit from a power source to the actuator in the system.

Electromechanical humidity control.   An electromechanical humidity controller is called a *humidistat* (Figure 4-4). A humidistat is an electrical switch that is actuated by changes in the amount of humidity in the air. A sensor in a humidistat senses changes in the amount of moisture in the air, and opens or closes an electrical switch when the moisture level deviates from a set point. Several different types of sensors are used. Some materials used as humidity sensors have the property of changing length when their moisture content changes. Other humidity sensors are electrical in nature and employ a device in which electrical resistance changes as the humidity level changes. In controllers using elements that change length, the elements grow shorter or longer as the moisture content changes, and this mechanical motion actuates an electrical switch in the control circuit.

A set-point indicator on the humidistat is set to the relative humidity desired for a space, and any deviation from that humidity will open or close the switch in the humidistat. The switch is wired in series in a control circuit running from the thermostat to a humidifier. If the humidistat calls for humidity, the humidifier is turned on, and if the humidistat is satisfied, the humidifier is turned off. A *humidifier* is a thermal or mechanical device that puts water vapor into the air.

Electromechanical airflow control.   Electromechanical airflow controls are electrical switches that are actuated by changes in air pressure, or

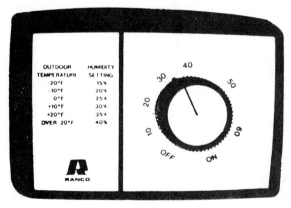

Figure 4-4   A controller for a system that humidifies or dehumidifies is called a humidistat.

airflow, and are used as operating controls or as proving switches. When used as operating controls, they control a system fan or duct dampers to control the flow of air into an area in a building to control building air pressure. When used as *proving switches* they are installed in fan outlets, ducts, or other air passages, to prove that air is flowing before some other action takes place.

Airflow proving switches are usually in the form of an electrical switch activated by a "sail" on an extension arm (Figure 4-5). The sail is inserted into a duct or other area in which air or other fluid flows. This type of switch is often called a *sail switch,* in reference to the sail on the arm. One application for a sail switch is for proving airflow when duct heaters are installed in an air duct with the blower located remotely from the heaters. The sail switch is used to "prove" that the air is flowing before the duct heater can come on. When used in this application, the contacts in the electrical switch in the sail switch are normally open (NO), and close only when the sail is depressed. The sail switch is wired in series in the control circuit to the duct heater, and the duct heater cannot come on unless the sail switch indicates an airflow.

Pneumatic Controllers

In pneumatic control systems the devices used as controllers are called *relays* or *controllers* (Figure 4-6). The function of a pneumatic controller is to regulate conditions in response to changes in a control variable. A pneumatic controller is in effect an adjustable pressure regulator. It works by regulating control system air pressure in relation to variations between the set point and controlled variable conditions. The set point is the condition desired in a space, and the controlled variable is the existing condition.

There are several different types of pneumatic controllers, with each type being designed for use in a specific application. A typical pneumatic operator has a flapper valve that opens or closes a small air vent. The flapper is controlled by an appropriate sensing element for the type of control desired. The flapper over the air vent controls the pressure in a chamber in the controller. Movement of a diaphragm in this chamber is affected by the pressure in the chamber. The

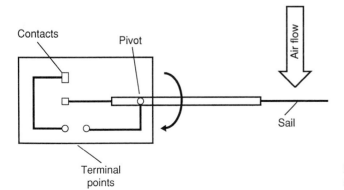

Figure 4-5 A sail switch is an airflow sensor and control.

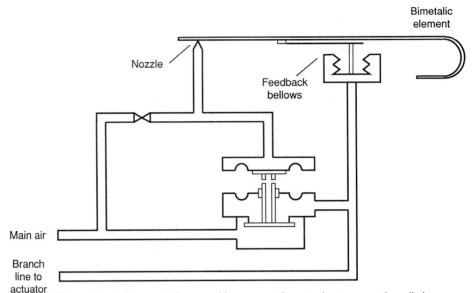

Figure 4-6 A thermostat for use with a pneumatic control system may be called a controller or relay.

diaphragm in turn is connected to a valve that controls the outlet control pressure from the controller to an actuator.

Pneumatic temperature control.   In a temperature controller used in a pneumatic control system, a bimetallic sensor controls the flapper valve that operates the controller. As the bimetallic element moves in response to changes in room temperature, the flapper allows more or less air to escape from the controller. This in turn controls the control air pressure out of the controller.

Pneumatic humidity control.   A humidity controller in a pneumatic system works basically the same as a temperature controller. The difference is that the flapper over the control vent is operated by a humidity-sensing element (Figure 4-7). If the humidity around the controller changes, the humidity-sensing element opens or closes the vent in the controller. Pressure within the controller chambers then controls the output control pressure.

Pneumatic airflow control.   An airflow controller in a pneumatic control system senses variations in air pressure. An airflow controller may be used in any one of several ways. The control may be used to sense air pressure in a part of a building in which positive ventilation is required, or it may be used as part of a combustion safety control system. In this case it senses air pressure at the combustion fan outlet to make sure that the combustion blower is running before the burner is energized.

The pressure tap of an airflow controller is connected to the source of pressure. This may be a fan outlet or, in a duct system, on an air duct. A small

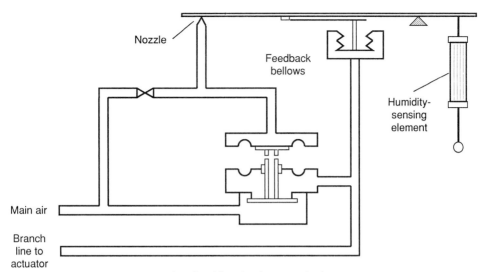

Figure 4-7 The sensor in a humidstat is a hygroscopic element.

air tube runs from the tap to the controller. The air tube is connected to a diaphragm or bellows in the controller. The diaphragm or bellows is connected by mechanical linkage to the flapper that controls air pressure in the controller, and consequently controls the outlet control pressure from the controller.

Electronic Controllers

A controller in an electrical control system is called a *sensor, transmitter,* or *operator.* Its purpose is to sense control variables and send an electronic signal to indicate variations between set point and controlled variable conditions. Various types of electronic controllers are used, depending on the application, to achieve the best control for each application. The sensor used in an electronic controller is one of several types of devices that will sense variations in the appropriate controlled variable.

An electronic controller is a combination of a sensing device and solid-state control devices (Figure 4-8). The solid-state control devices are part of a prewired circuit board. The sensing device will sense variations in the controlled medium, while the electronic devices convert a signal related to the variations to a low-voltage direct-current (dc) signal. The signal is in a form that can be further processed by a microprocessor. The microprocessor is located either in the controller or in a central processing unit.

Electronic temperature control. The controller used in an electronic temperature control system senses variations between a set-point temperature and controlled variable temperature. Several different types of sensors are used in electronic thermostats. Among these are thermistors, potentiometers, and

Controllers **53**

Figure 4-8 A thermostat used in an electronic control system is made up of solid-state control devices.

resistors. The most commonly used sensor is a thermistor. A *thermistor* is a sensor in which electrical resistance changes as the temperature changes. In an electronic thermostat the thermistor is wired in series in a sensing circuit and located in the area in which the temperature is to be controlled. If the temperature around the thermistor changes, the electrical resistance changes. These changes are converted to changes in the electrical current flowing through the resistor and through the control circuit. The change in electrical current is proportional to changes in the temperature. The current is then used then to determine variations from set-point temperature. The electrical signal goes to a central control unit, where it is converted to an output signal that can operate an actuator in the control system.

Electronic humidity control. The controller used in an electronic humidity control system has a sensor that is sensitive to moisture in the air (Figure 4-9). The sensor is wired into an electrical bridge in which the variations in humidity are translated to electrical current variations. The current variations are then sent to a central processing unit, where they are converted to output signals to the actuator in the system.

Electronic airflow control. An electronic controller used for sensing air pressure uses a solid-state device called a *piezoelectric crystal* (Figure 4-10). In a piezoelectric crystal, pressure changes cause a change in electrical resistance. A capillary tube is connected to the area in which the pressure is to be controlled and runs to the controller. In the controller the tube is connected to a diaphragm or bellows element. The diaphragm or bellows is connected by

Figure 4-9 The sensor in an electronic humidstat senses humidity electronically.

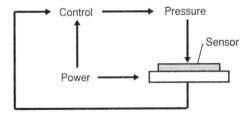

Figure 4-10 An electronic pressure sensor typically utilizes a sensing element called a piezoelectric crystal.

mechanical linkage so that if the air pressure changes, pressure will change on the piezoelectric crystal. The crystal is connected in series as a variable resistor in a bridge circuit. As the pressure read by the control varies, the variations are translated through a control panel as output to an actuator.

SIGNAL

The *signal* in a control system is the medium used to relay information from a controller to a control center or actuator, indicating a need for action (Figure 4-11). The signal medium used in an electromechanical system is relatively low-voltage ac electricity, but the signal itself is the presence of voltage or the absence of voltage in the electrical control circuit. The signal medium used in

Signal

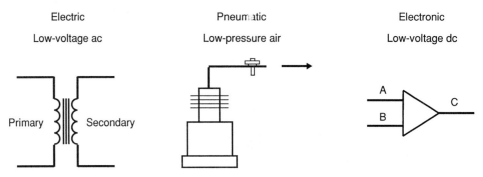

Figure 4-11 The basic control systems are electrical, pneumatic, and electronic.

a pneumatic system is low-pressure air, and the signal is a variation in that pressure. The signal medium used in electronic control systems is low-voltage dc electricity, while the signal itself is voltage variations in the electrical circuit.

ACTUATORS

Actuators are the components in a control system that turn equipment on and off or otherwise regulate its operation. Actuators in a control system function as they receive a control signal from a controller or some intermediate control between the controller and the actuator. Many different types of actuators are used, depending on the type of control system used and also depending on the type of equipment to be controlled.

Electromechanical Actuators

Actuators in electromechanical control systems are usually some type of electrical switch or operator. Some of the common actuators are relays, contactors, magnetic starters, and solenoids. In modulating electromechanical control systems damper and valve motors are often used.

Relays. *Relays* are electrical control switches in which one circuit controls another (Figure 4-12). They are often used in electrical control systems in which a control circuit operates a power circuit. Quite often a low-voltage circuit controls a high-voltage circuit through a relay. A relay typically has an electromagnetic coil that is energized by the control circuit (Figure 4-13). The coil operates a set of contacts in a secondary circuit. The contacts either open or close when the coil is energized, depending on the type of relay used. In some relays the contacts are normally open (NO) and are pulled in when the coil is energized. In some relays the contacts are normally closed (NC) and open when the coil is energized.

Figure 4-12 A magnetic relay allows the control of one electrical circuit by another.

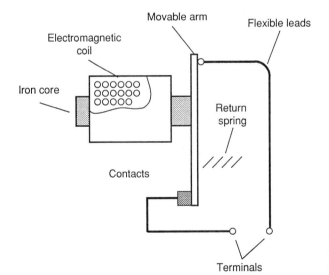

Figure 4-13 In a magnetic relay a magnetic coil actuates the contacts to make or break contact.

Contactors. A *contactor* is an electrical device similar to a relay but designed so that it can carry heavier current loads (Figure 4-14). Contactors are used to control electrical circuits in which the amperage is higher than a relay can handle. A contactor has an electromagnetic coil that is energized by a control circuit. The magnetic field of the coil moves an armature back and forth. The armature has a set of contacts on it, and a matching set of contacts

Actuators   **57**

Figure 4-14 A magnetic contactor is used to control line circuits to larger motors.

is mounted on the frame of the contactor. As the armature moves, the contacts open or close, either making or breaking a circuit wired through them. The power circuit of the system is wired through the contacts.

Magnetic starters. A *magnetic starter* is similar in construction to a contactor, but it has overload protection built in for the electrical power circuit that it controls (Figure 4-15). Magnetic starters are usually used as control devices when large motors are used on a system. They should be used when the motors do not have built-in overload protection.

In a magnetic starter, an electromagnetic coil actuates a set of contacts, as in a contactor or relay. The contacts open and close as the coil is energized or deenergized, but in addition, electrical heaters are located where they sense the electrical current in the electrical power circuit. A second set of normally closed contacts are located in the control circuit wiring. If an overcurrent con-

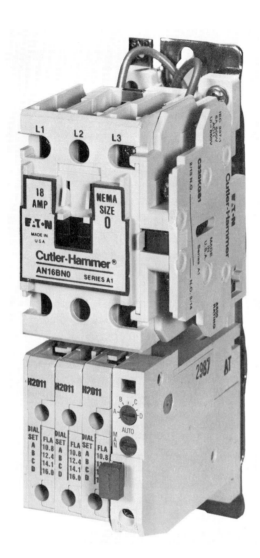

Figure 4-15 A magnetic starter is a motor controller with overload relays in it.

dition occurs in the power circuit, the heaters open the contacts in the control circuit and the main contacts in the starter are opened, shutting the equipment that the contactor is controlling off. The heaters for a contactor come in different ratings to protect against a number of different amperages. They can be selected for each job as needed.

Solenoids. A *solenoid* is an electrically controlled device in which a control circuit can be used to open or close switches, valves, or other control devices (Figure 4-16). A solenoid has an electromagnetic coil with an open core. A movable iron slug moves up or down in the core. When the solenoid coil is energized, the slug moves up in the core, and when it is deenergized, it

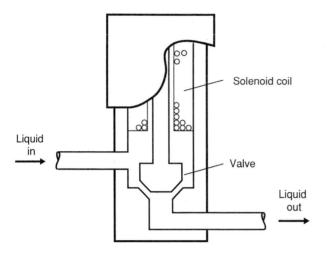

Liquid
in

Solenoid coil

Valve

Liquid
out

Figure 4-16 A solenoid operator utilizes a magnetic coil with a movable core to open or close a valve or contacts.

drops down. The slug is connected by mechanical means to the part of the device that is to be activated. If the relay operates an electrical switch, the connection is to a set of contacts. If the solenoid operates a valve, the connection is to a valve operator.

Pneumatic Actuators

The most commonly used actuators in pneumatic systems are either diaphragm or piston operated. They are generally used to operate valves or dampers. If the system is a combination electromechanical and pneumatic system, the actuator may operate an electrical switch. A *diaphragm actuator* is a device in which a diaphragm is contained in a housing (Figure 4-17). A diaphragm is a thin metal or composition membrane fastened between two hollow chambers. An air line from a controller runs to one of the chambers. As air pressure from the controller varies, the diaphragm moves back and forth. The diaphragm is connected to an actuator by a mechanical linkage through a lever connection. The linkage operates dampers or a valve.

Diaphragm actuators are important devices for opening and closing large valves and dampers. They can be built as large as necessary. Pressure exerted by the diaphragm is a function of the area of the diaphragm and the pressure exerted. The size of the actuator is related to the pressure exerted. As the area of the diaphragm increases, pressure conveyed by a relatively small amount of control pressure is multiplied by the area of the diaphragm used.

A *piston actuator* is a piston that moves back and forth within a cylinder (Figure 4-18). Control air pressure is introduced in the cylinder at one end and there is a compression spring in the other end. If the control air pressure modulates, the piston moves back and forth with it. A connecting rod runs from the piston to a damper or valve operator. As the piston moves, so does the damper or valve operator. When controlled by air pressure from a pneumatic

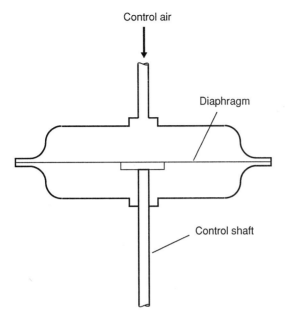

Control air

Diaphragm

Control shaft

Figure 4-17 Power developed by a diaphragm operator is proportional to the area of the diaphragm.

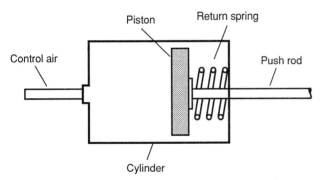

Piston Return spring

Control air Push rod

Cylinder

Figure 4-18 A piston operator can be used to operate dampers where a reasonably large lateral motion is required.

controller, diaphragm and piston actuators are inherently modulating control devices.

A *pneumatic–electric switch* is a control device in which a pneumatic control signal is transposed to an electrical signal (Figure 4-19). In some applications it is desirable to have a pneumatic control signal operate an electrical device. In this case a pneumatic–electric switch is used. There are also *electric–pneumatic switches,* in which an electrical signal is transposed to a pneumatic signal. This control is used to control pneumatic devices when an electrical signal is used.

Electronic Actuators

Electronic actuators are usually called controlled devices (Figure 4-20). These may be sequencers, proportional controllers, transducers, damper actuators,

Actuators

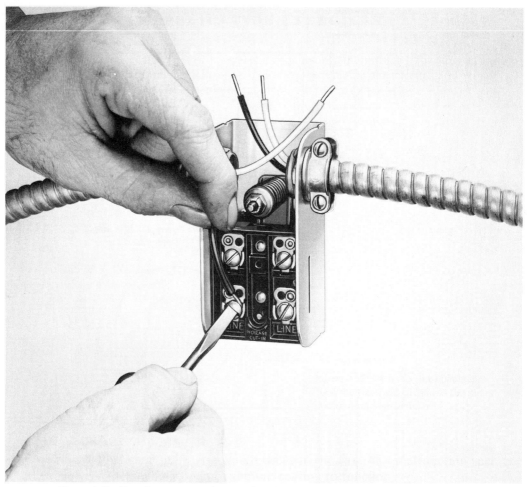

Figure 4-19 By using a pneumatic–electric switch, a pneumatic control system can be interfaced with an electric system.

valve motors, or relays. The low-voltage signal used in electronic systems is generally too weak to operate an actuator directly, so an adapter is used between the controller and the actuator which interprets the control signal and sends a stronger ac signal to operate the controlled device.

Most actuators used in electronic control systems are similar to those used in electromechanical systems. There are some variations in sequencers and certain other devices where electronic control devices can be used more efficiently than can electromechanical systems.

One special type of actuator used in electronic control systems is a self-contained hydraulic operator. This operator is used as a damper or valve operator. The self-contained hydraulic operator contains a small hydraulic motor and pump, an oil reservoir, a hydraulic valve, and a piston operator. The motor

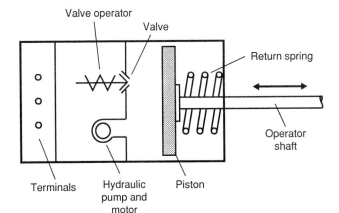

Figure 4-20 An electronic hydraulic operator is an operating device in which an electronic signal can be used directly to operate a control device.

and pump run continuously. The valve is controlled by a control signal from the control panel. When the valve is opened on a call from the panel, oil is pumped into the piston cylinder. This causes the piston to move out in the cylinder. When the signal calls for control action to stop, the valve closes and a relief valve opens to allow the oil in the cylinder to flow back into the reservoir. This causes the piston to move back to its original position.

FEEDBACK

Feedback is a signal sent back to a controller from the controlled variable indicating that conditions are normal. In an electromechanical control system, feedback is the absence of voltage in the control circuit when the control signal is discontinued. When a condition calling for operation of the controller is corrected, the sensor in the controller opens the electrical switch in the control circuit and the actuator returns to normal operation.

Pneumatic feedback is a change in air pressure in the pneumatic control system. This change repositions the actuator to a satisfied position. *Electronic control system feedback* is accomplished by changing voltages in the low-voltage ac electrical signal. When a condition is satisfied, the low-voltage signal from the controller is modified and the actuator returns the operating equipment to a satisfied position.

SUMMARY

The basic elements of a control system are the devices used to sense conditions, produce a signal to indicate those conditions, and produce results to correct deviations from the norm in the conditions, and a way to interpret controlled conditions to indicate when corrections have been made. The four basic elements are sensing, signal, control, and feedback.

Summary

The three main types of control systems are electromechanical, pneumatic, and electronic. Because of variations in the three types of control systems, the basic elements are sometimes called by different names, but they serve the same functions.

The two main parts of any system are the controllers and actuators. Controllers are the part of the system in which sensors of some kind sense differences between set-point conditions and controlled variable conditions. Actuators are that part of the system in which a signal received from the controller initiates some action to correct the differences.

QUESTIONS

4-1. Name the four main elements of a control system.

4-2. What is the primary function of the element in a control system that is called by the generic term *controller?*

4-3. *True or false:* All of the elements of a control system are called by the same names in each type of control system.

4-4. *True or false.* Thermostat, relay, and controller are names given to the same element of a control system in different types of control systems.

4-5. What is the name of the control device used as a controller in an electromechanical temperature control system?

4-6. What is the most obvious application for a remote-bulb thermostat?

4-7. *True or false:* A humidistat is a device that provides moisture to a building when the humidity in the building is too low.

4-8. Name two uses for airflow control devices.

4-9. How does a pneumatic controller function as a variable-pressure regulator?

4-10. What are three names used for the control device that functions as a controller in an electronic control system?

4-11. What is the name of the most commonly used sensor for sensing temperature in an electronic control system?

4-12. *True or false:* A control device called a piezoelectric crystal is used for sensing pressure in an electronic airflow control system.

4-13. Name the three signal media used in electric, pneumatic, and electronic control systems.

4-14. What is the main function of the element of a control system known generically as an actuator?

4-15. Name four different control devices used as actuators in an electromechanical control system.

4-16. What is the name of one electrical control device that is used primarily to control one electrical circuit with another.

4-17. Name two types of actuators used in pneumatic control systems.

4-18. *True or false:* Because of their size, diaphragm actuators are useful only to control. small valves.

4-19. What is another name used to identify actuators used in electronic control systems?

4-20. Name one special type of actuator used with an electronic control system in which a hydraulic motor and pump are involved.

4-21. What is *feedback* in respect to each of the three main types of control systems?
   A. Electric   _____
   B. Pneumatic   _____
   C. Electronic   _____

APPLICATION EXERCISES

4-1. Fill in the blanks on the accompanying diagram that shows the main parts of an electromechanical control system. Describe the function of each part in the system.

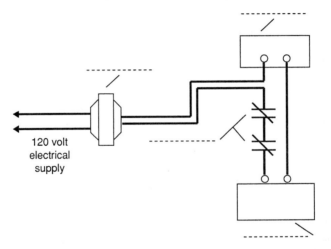

4-2. Fill in the blanks on the accompanying diagram that shows the main parts of a sail switch. Describe the function of each of the parts.

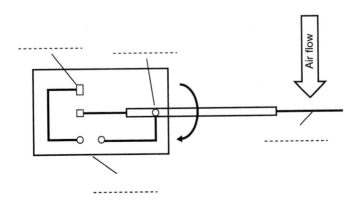

4-3. Fill in the blanks on the accompanying diagram that shows the main parts of a pneumatic temperature controller. Describe the function of each of the parts.

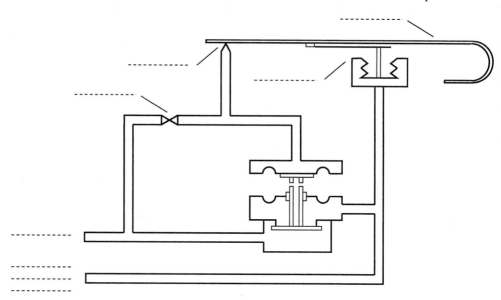

4-4. Fill in the blanks on the accompanying diagram that shows an electric control system with a step controller used on an electric furnace. Describe the function of each of the parts.

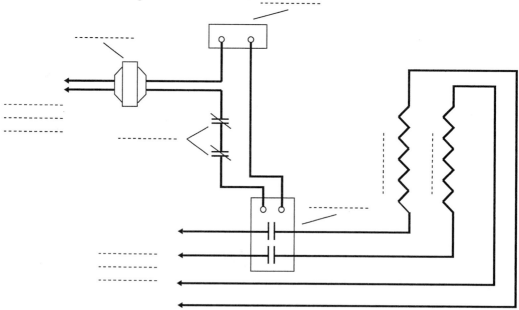

4-5. For this excercise use an operative gas or oil furnace, trainer or facsimile, with the control wiring removed. Draw a control wiring diagram for the unit showing the controls to be used and the connections for an electric control system. Wire up the unit completely for electric control system operation. Have it checked by your instructor, and then operate the system.

4-6. For this exercise use an operative hot water or chilled water coil, trainer or facsimile, with the controls removed. Draw a diagram of a pneumatic control system to operate the coil, then install the complete pneumatic control system on the coil. Have the diagram and the system checked by your instructor, and then operate the system.

5

Control Function

A comfortable indoor climate is one in which the temperature, humidity, and airflow in the spaces in a building are controlled so that they do not cause any feeling of discomfort to an occupant. Control of these variables constitutes a comfort control system. This control is achieved by the use of various devices that operate and regulate the operation of the heating, cooling, and ventilation equipment in the building. Each control device is designed to provide a certain function in the control system. The main function of the devices is to sense conditions, send a signal related to the condition, and provide an action to control the condition.

SENSING CONDITIONS

A general term for the device used to sense conditions, and usually to send a signal to an actuator related to those conditions, is *controller*. In an electro-mechanical system a controller is usually called a *thermostat, humidistat,* or *pressurestat;* in a pneumatic system a controller is called a *relay;* and in an electronic system it is called a *controller.*

Temperature Sensing

Control of the temperature in a building is perhaps the most important element of comfort (Figure 5-1). Most temperature controllers have both a set-point indicator and a sensor. The set-point indicator is a pointer that can be moved across a scale with temperature readings on it. The indicator is "set" to the temperature desired on the scale. The sensor is a device that "senses" the actual temperature at the controller. A controller originates a control signal relative to differences between the set point and actual conditions sensed at the controller.

A *sensor* is the part of the controller that actually monitors the temperature and indicates variations from the set point that require correction by the heating or cooling equipment. Several different types of sensors are used. Some of the more common ones are bimetallic elements, pressure-spring elements, thermocouples, electric resistance, and various electronic sensors.

Bimetallic sensor.   A *bimetallic sensor,* the most common type of sensor used with temperature-sensing devices, is made up of two types of metals, each with a different coefficient of expansion (Figure 5-2). The two types of metal are fastened together so that as they are cooled or heated, they warp in

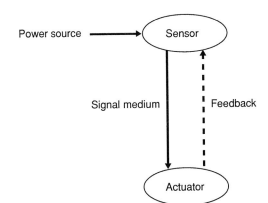

Figure 5-1   Four elements of a control system are the sensor, a signal, an actuator, and a feedback signal.

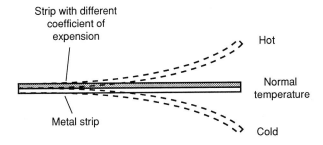

Figure 5-2   A bimetallic element is often used as a temperature-sensing device.

one direction or another. The warping, or bending, action is used to open or close a switch, or to operate another control.

Bimetallic sensors are made in a number of different shapes. Some of the common ones are leaf, spiral wound, and helical. Leaf and spiral-wound bimetallic sensors are commonly used in thermostats. Helical bimetallic sensors are often used in fan controls, limit switches, and some combustion control devices. A leaf-type bimetallic sensor is one in which the metals are straight and simply bend back and forth in one plane when they are heated or cooled. Electrical contacts are opened and closed as the bimetallic sensor moves back and forth.

A spiral-wound bimetallic element is normally used in a mercury bulb controller. A spiral-wound bimetallic element is fixed to a connection from the set-point indicator at the center point. Heating or cooling causes the bimetallic sensor to rotate around its center. A mercury bulb switch is mounted on the top of the bimetallic element (Figure 5-3). The rotary motion of the bimetal sensor causes the mercury bulb to move back and forth over its center of gravity. A mercury bulb switch is a glass tube that is sealed at both ends. A drop of mercury is sealed in the tube. A slight bend in the tube causes the mercury drop to remain in one end or the other when the tube is level through its length. Electrical leads are sealed in the tube at one end. If the tube tips toward that end, the mercury drop makes contact between the leads. If the tube tips in the opposite direction, the contact is broken. The leads in the bulb are connected by flexible wires in the control circuit of the thermostat. The contacts in the bulb are opened and closed as the bimetallic sensor rotates.

Pressure-spring sensor. *Pressure-spring sensors,* often used in remote-bulb thermostats, use pressure that is generated in a closed system to move a diaphragm or bellows operator (Figure 5-4). The motion generated by the diaphragm or bellows opens or closes a switch. The control has a small bulb filled with a volatile liquid attached to the diaphragm or bellows operator by a small capillary tube. When the liquid in the bulb is heated or cooled, it expands or contracts and moves the diaphragm or bellows by internal pressure. A mechanical connection from the diaphragm or bellows operates a switch. Remote bulb thermostats are used when the temperature is measured at a point

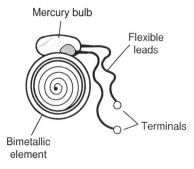

Figure 5-3 A bimetallic element is often combined with a mercury bulb switch to make a temperature-sensing electrical control element.

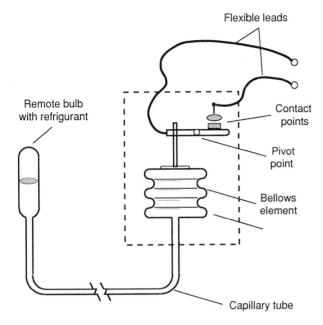

Flexible leads

Remote bulb
with refrigurant

Contact
points

Pivot
point

Bellows
element

Capillary tube

Figure 5-4 A remote-bulb control usually uses a diaphragm or bellows element to provide mechanical motion to actuate a switch.

at which it would be difficult to locate a thermostat. Temperature readings on heating elements or cooling coils inside a cabinet are often taken with remote-bulb thermostats.

Thermocouple. A *thermocouple* is a device that produces an electrical potential, or voltage difference, when it is heated (Figure 5-5). A thermocouple is constructed of two different types of metal wires twisted together at one end. This twisted section is called a *couple*. When the couple is heated a potential difference is generated across the other ends of the wires. The amount of potential difference is proportional to the temperature of the couple, so the difference is used to indicate temperature on a scale calibrated to show temperature instead of voltage.

The electrical output of a thermocouple is a function of the heat applied to the couple and to the materials used in the wires. Different combinations of wires produce different voltages and thence different temperature readings. The wires used in the thermocouple itself are often platinum and some other

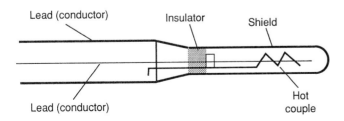

Lead (conductor)

Insulator

Shield

Lead (conductor)

Hot
couple

Figure 5-5 A thermocouple is used to sense a flame in a pilot safety combustion control system.

Sensing Conditions

wire. Since platinum is very expensive, the lead wires from the couple to the indicating instrument are usually copper or some other less expensive wire.

Electric resistance sensor. An *electrical resistance sensor* is a wire-wound resistance element that is used to measure temperature (Figure 5-6). The electrical resistance of a wire is directly affected by the temperature of the wire. An electrical resistance sensor is a wire coil. The coil is wired in series in an electrical circuit in which a current of electricity is flowing. As the electrical resistance of the wire in the coil varies with the temperature around the coil, the current flowing in the circuit varies. The coil circuit is wired into a bridge circuit as a variable resistor. The bridge circuit is used in an electrical circuit that shows the current variations as temperature variations on a readout device.

A *bridge circuit* is an arrangement of four resistors in two parallel circuits (Figure 5-7). There are two resistors in each of the parallel legs. One of the resistors is a variable resistor controlled by a sensor in a controller. In this case the variable resistor is the sensing coil. The other three resistors are rated at

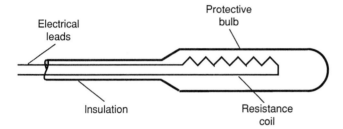

Figure 5-6 A wire-wound resistance sensor is often used to sense temperature differences.

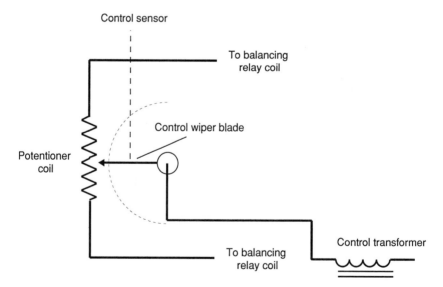

Figure 5-7 A potentiometer can be used to divide an electrical signal proportional to a sensed condition.

the same resistance value. The bridge circuit is connected to a pair of conductors leading to an ac electrical power source at the junctions of the two parallel circuits. An electrical circuit leading from the junctions between the two resistors in each parallel leg of the circuit provides an output signal.

At set-point conditions the variable resistor has the same resistance value as the other three resistors in the circuits. The current flowing in the two parallel circuits is the same, and the output from the circuits indicates no current flow. If the electrical resistance in the wire coil sensor changes, the electrical output from the output terminals changes, and this is shown on a meter connected in the circuit. The meter is calibrated in degrees Fahrenheit or Celsius rather than in milliamperes.

The sensing element of an electrical resistance thermometer is usually contained in a metal bulb. The wire-wound resistor is enclosed in metal, and wires run from the bulb to the bridge circuit. Various materials are used for the wire in the sensing element, depending on the range of temperatures to be measured (Figure 5-8). Platinum wire is used in an electrical resistance element if the temperature range is -40 to $+120°F$. Nickel wire is used over the range -250 to $+600°F$. Copper wire is used for a range from -328 to $+250°F$. Some types of radiant heat sensors are also used as temperature-sensing devices. These are usually used in conjunction with electronic circuitry and electronic control devices.

Humidity Sensing

Humidity is water vapor held in suspension in the air. Two terms are used to describe the amount of water vapor per unit volume of air. These are relative humidity and humidity ratio. *Relative humidity* is the percentage of water in the air compared to what there would be if the air were saturated. The *humidity ratio* is the actual amount of water vapor, in grains or pounds (lb) of water

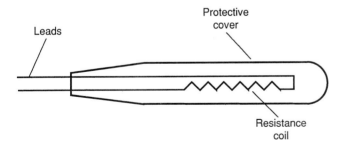

Resistance Values

Platinum: −40° to 120 °F, 25 ohms at 32 °F
Nickel: −25° to 60 °F, 100 ohms at 77 °F
Copper: −328 to 250 °F, 10 ohms at 77 °F

Figure 5-8 Different materials can be used in resistance elements to function within different temperature limits.

Sensing Conditions

vapor per pound of dry air. A grain is 1/7000th of a pound. Two different types of instruments are used for determining humidity: hygrometers and psychrometers. A *hygrometer* is a device that measures humidity by sensing physical or electrical changes that occur in various materials as they absorb moisture. *Psychrometers* register temperature differences that occur between actual conditions and saturated conditions.

Hygrometric sensor. *Hygrometric sensors* that depend on physical changes, employ animal hair, human hair, animal membrane, organic membranes, or other materials that change length as their moisture content changes (Figure 5-9). Their moisture content is actually a function of the moisture content of the air surrounding them. The change in element length is transmitted by mechanical means to an electrical switch or recorder.

Electric sensor. Another type of hygrometric sensor is an *electrical humidity sensor.* In an electrical humidity sensor a transducer converts humidity changes into electrical resistance changes. A transducer is a control device that changes one form of energy to another. The transducer used as a sensor in a humidistat is wired into a low-voltage circuit that transmits changing voltages to an indicating or recording device. Another type of electrical humidity sensor uses a capacitance probe to sense humidity conditions. Changing humidity causes changes in the capacitance of the probe, and these changes are used to indicate the humidity level.

Psychrometric sensor. *Psychrometric sensors* used for determining humidity levels are usually a combination of dry-bulb and wet-bulb thermometers. The bulb of the wet-bulb thermometer is artificially wetted with water while readings are taken. The difference between the dry-bulb reading and the wet-bulb reading indicates the amount of water vapor in the air. The psychrometer commonly used by technicians is called a *sling psychrometer,* two ther-

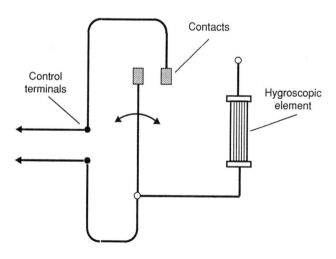

Figure 5-9 A humidistat is used to sense space humidity in a humidity control system.

Control Function   Chap. 5

mometers mounted on a common base with an arrangement for keeping the bulb of one of the thermometers wet with water (Figure 5-10). In use the psychrometer is whirled in the air to cause a constant flow of air to pass over the thermometer bulbs. Water evaporating from the wetted bulb cools the bulb slightly and gives a reduced temperature reading. The difference between the wet-bulb and dry-bulb reading is called the *wet-bulb depression*. This is a function of the amount of moisture in the air, since the rate of evaporation is a function of the same.

Many indicating and recording humidifers use the same principle as the sling psychrometer, the difference being that they use electrical temperature sensors, with one sensor being dry and the other wetted. The humidity is still indicated as a function of the temperature difference.

Airflow Sensing

Airflow within the conditioned spaces in a building is an important element in providing a comfortable climate. Constantly moving air prevents stagnation and reduces stratification. Moving air can be filtered, and ventilation air can easily be introduced to the spaces with the moving air.

Fans. Generally, airflow is provided by a fan that runs when either heating or air conditioning is called for, or in the case of commerical buildings, runs all of the time that a building is occupied. An automatic fan control switch turns the fan motor off and on, on demand for heating or cooling, or a blower control switch on the thermostat is set for continuous operation.

Filters. An air-filtering system is used to clean the air as it circulates through the ductwork in a building. Filters are installed in the return-air side of a duct system. Filter systems in smaller buildings are usually self-contained and controls are not involved. On some larger filter systems manometers may be used to indicate how dirty the filters are by showing air pressure drop across the filter bank (Figure 5-11). A *manometer* is a device that registers air pressure in a duct system. When manometers are used to indicate filter loading with dirt or dust, one is placed on the upstream side of the filter and another downstream. The difference between the two readings is an indication of resistance to airflow across the filters. In some sophisticated filter systems roll filters may be used, in which the filters advance automatically as they become dirty. The controls to advance the roll filters consist of manometers to measure pressure drop across the filter bank and relays to control a motor to advance the filter.

Ventilation. Ventilation air is provided to a building by the introduction of outside air. Dampers are used to control the amount of air introduced (Figure 5-12). Some ventilation air dampers are manually controlled and others are

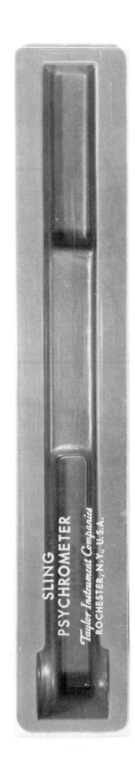

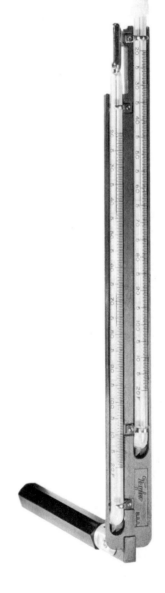

Figure 5-10 A sling psychrometer is used to measure the difference between wet- and dry-bulb temperature for determining relative humidity.

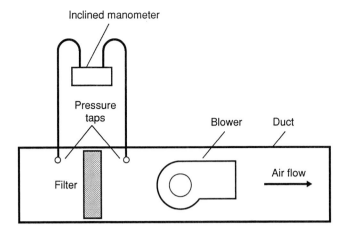

Figure 5-11 Filter loading with contaminants can be measured by using a manometer to determine pressure drop through the filter.

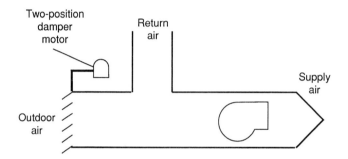

Figure 5-12 Outdoor air dampers are often controlled by use of a damper motor.

controlled by damper motors. When motorized dampers are used they are usually controlled by temperature or pressure sensors. Some ventilation air is required whenever spaces in a building are occupied. The amount of air required is determined from the number of people present, activity of the people, and the amount of contamination of the air by smoking or other forms of pollution. Modulating outside-air dampers are arranged so they cannot close beyond a point where the minimum amount of air required in a space is admitted.

SIGNALS

Signals are the means used to transfer data from a controller to an actuator in a control system. The controller may be a thermostat, humidistat, pressurestat, or any other device used for indicating the difference between the set point and the controlled variable. Commonly used media for signals are line or low-voltage ac electricity, low-pressure air, and low-voltage dc electricity.

Line or low-voltage electrical ac signals are used in electromechanical control systems. If line voltage is used, it will usually be the nominal voltage used for the system being controlled. When low voltage is used it is normally 24 volts. An electrical control system is basically a digital or off–on system (Figure 5-13). A control circuit consists of a power source, conductors, a controller that senses conditions and functions as a switch, and an actuator. When the controller senses a predetermined difference between the set point and the controlled medium, an electrical switch closes and the circuit is energized. This causes the actuator to function and action is taken to correct the difference.

Low air pressure is used as a signal medium in a pneumatic control system. A typical system works with a maximum pressure of about 20 psi and working pressures between 3 and 18 psi. A pneumatic control system is basically a modulating or analog system. Changing conditions sensed by the controller are converted to changing air pressure in the system, and the actuator functions in response to the changing pressure. A pneumatic control system can be interfaced with an electrical or electronic system by the use of switches in which air pressure changes are used to open or close electrical circuits (Figure 5-14).

A low-voltage dc electrical signal is used in an electronic control system. A typical system works within a range of about 5 to 15 volts. In an electronic control system a central control panel is used in which the supplied ac electricity goes through a transformer where the voltage is lowered. It also goes through a rectifier where it is changed from an ac to a dc signal. This low-voltage dc signal is sent out to controllers in the system. In the controllers, differences between the set-point conditions and the controlled-medium conditions are sensed by an electronic sensor. Any variations are reflected in changes in voltage in the signal that is then sent back to the control center. In the control center the changing signal is converted to an outgoing electrical signal sent to an actuator, which controls the operation of the equipment in the system (Figure 5-15).

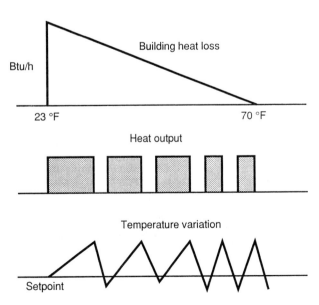

Figure 5-13 A control system that uses electricity as a control medium is generally a digital system, providing on–off control.

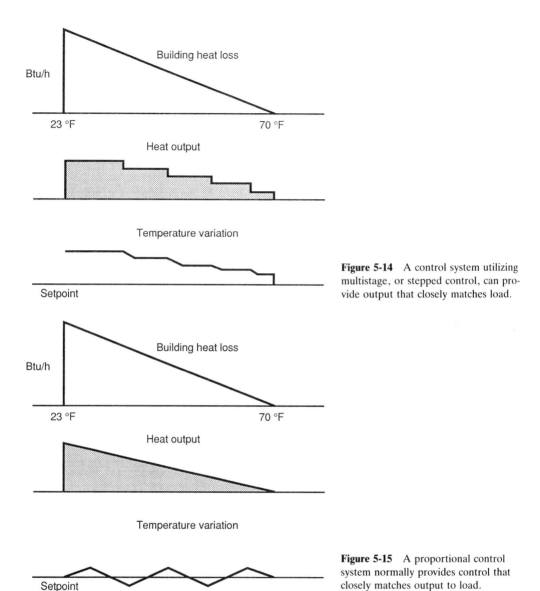

Figure 5-14 A control system utilizing multistage, or stepped control, can provide output that closely matches load.

Figure 5-15 A proportional control system normally provides control that closely matches output to load.

ACTUATORS

The *actuator* in a control system is the device that receives a signal from a controller and performs come action to correct the difference between set-point and control-point conditions. In electromechanical systems actuators may be relays, solenoids, starters, contactors, or other controls that turn some piece of equipment on or off. In a pneumatic system the actuator may be a piston operator, diaphragm, or bellows-type operator. In an electronic control system the actuator will usually be similar to those used in electromechanical systems.

In general, there are three different types of actions provided by typical actuators: on–off (digital), stepped (sequenced), and proportional (analog) or floating.

On–off control. *On–off* or *digital control* is a natural function of an electromechanical control system. When a switch closes in an electrical circuit an electric current flows through the conductors in the circuit. This is an on signal. When the switch opens and no current flows, this is an off signal. A digital actuator can be any electrical device that is actuated when an electrical circuit is energized and deactivated when the circuit is deenergized. Activation of a relay by an electrical signal can start an electric motor, open a valve, or do any one of many other things in an electrical control circuit. Deenergization of the relay will stop the motor, close the valve, or stop any of the other actions.

Stepped control. *Stepped control* is that used when several different digital controls are staged on or off in timed intervals. Stepped control is often used for controlling larger electrical loads (Figure 5-16). Electrical heaters can be made up of several electrical resistance heaters, each called an *element*. It is not desirable to have all the elements come on at once because of the large inrush current when this happens. If the elements are brought on one at a time by stepped control, the inrush load is spread out over time. The heat output from the elements can also be controlled to more closely match a heating load.

Stepped control is achieved by using a controller that mechanically turns a series of switches on or off in sequence. This type of controller is often called a *sequencer*. Some step controllers have a series of contacts that are opened and closed when a bimetallic element is heated by an electrical heater. When the control circuit is energized, the contacts close in sequence, and when the control circuit is deenergized, the contacts open in sequence.

Proportional control. *Proportional control* is control in which response to a changing condition is proportional to the changes. It is analog control. In proportional control, if the controlled variable condition varies by a given amount from the control-point condition, a change is made by the system actuator in the same proportion. Proportional control in an electromechanical control system is achieved by using controllers and actuators with

Figure 1

Left   Right

Figure 5-16 A stepped-controller, or sequencer, is used to provide multistage control.

matching components of a bridge circuit. But it is inherent, or natural, in both pneumatic and electronic systems, as a basic function of the system.

FEEDBACK

Feedback of information is an important function of a control system. *Feedback* is information related to the completion of a task. A feedback signal is most often generated in the controller by the sensor in a control system. While a sensor in a controller indicates a variation between set-back conditions and ambient conditions of a controlled medium to operate an actuator, it will usually also indicate when the condition is corrected by sending another signal. This is called feedback.

Electromechanical, pneumatic, and electronic control systems all require that some form of feedback occur. Because the signal for each type of control is different, so is the feedback signal. In a digital system, which in most cases is related to an electromechanical control system, the closing of an electrical circuit, and subsequent flow of electrical current, is a signal that some control is desired. The opening of the circuit, and subsequent interruption of the current flow, is a feedback signal that the control condition is satisfied. In a proportional system, related to both pneumatic and electronic control systems, feedback is in the form of variation in the signal sent from the controller to the actuator. In a pneumatic system the variation is in the form of a change in the low-pressure control air. Usually, a lower pressure indicates that a condition is satisifed. In an electronic control system the variation is in the form of a reduction in the dc voltage originated by the controller.

SUMMARY

Control system function is the operation of the complete control system, including all the control devices and components. It includes sensing conditions, signals, actions, and feedback.

Controllers are those devices that sense differences between set-point conditions and controlled-medium conditions. All controllers contain some type of sensor that indicates this difference and originates a signal to indicate the difference.

Signals differ depending on the type of control system used. Electromechanical control systems use low-voltage ac electricity as a signal. Pneumatic control systems use low-pressure air as a signal, and electronic control systems use low-voltage dc electricity as a signal. The signal of a control system causes an actuator of some kind to produce some action to correct the difference sensed by the controller. Feedback is a signal indicating that any difference between the set-point condition and controlled-medium condition has been eliminated.

QUESTIONS

5-1. What is the name of the control used to sense temperature and originate a control signal in an electric control system?

5-2. What is the name of the part of a thermostat that actually senses the temperature of the air?

5-3. Describe a bimetallic sensor.

5-4. Name three shapes commonly used for bimetallic sensors.

5-5. Describe a mercury bulb switch.

5-6. Describe a pressure-spring sensor used in a remote-bulb thermostat.

5-7. Name two types of instruments used to sense humidity.

5-8. What is the function of a transducer as used in a humidstat?

5-9. How are outside-air dampers controlled to make sure that the minimum of air required for ventilation is always brought into a building?

5-10. What signal medium is used in an electric control system?

5-11. What four elements constitute a control circuit?

5-12. What signal medium is used in a pneumatic control system?

5-13. What signal medium is used in an electronic control system?

5-14. Name the three different types of actions provided by actuators.

5-15. Match the term in the first column below with the phrase that best matches it in the second column by placing the letter that precedes the description in the space provided preceding the term.

A. _____ Digital	a. stepped controller
B. _____ Stepped	b. inherent in pneumatic system
C. _____ Proportional	c. deactivated is "off"

5-16. Match the term in the first column below with the phrase that best matches it in the second column by placing the letter that precedes the description in the space provided preceding the term.

A. _____ Digital	a. no voltage
B. _____ Stepped	b. steps off
C. _____ Proportional	c. varies with variable

APPLICATION EXERCISES

5-1. Using a low voltage power source, such as a transformer, a low voltage thermostat and a low voltage light bulb, wire them up in a control circuit in which the thermostat controls the on–off cycles of the light bulb. Turn the light bulb on and off by adjusting the set point indicator of the thermostat below and above the ambient temperature. Describe the operation of the system and compare the components with the control components in an electromechanical control system.

5-2. Name the parts of an electromechanical thermostat on the accompanying line drawing of a typical thermostat, by writing them in the spaces provided.

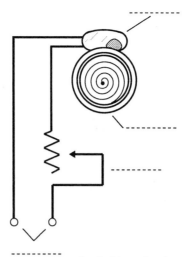

5-3. On the accompanying diagrams of a bridge circuit controlling a damper motor, show which way the motor will turn as the potential wiper blade moves as shown on the diagrams. Show which way the balancing damper arm moves in each case, and which side of the circuit has the higher amperage.

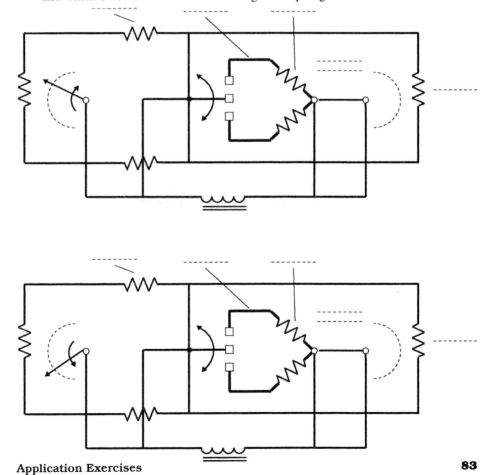

5-4. Using a low voltage power source, such as a transformer, a low voltage humidistat and a low voltage light bulb, wire them up in a control circuit in which the humidistat controls the on–off cycles of the light bulb. Turn the light bulb on and off by adjusting the set point indicator of the humidistat below and above the ambient humidity of the room. Describe the operation of the system and compare the components with the control components in an electromechanical control system controlling humidity.

5-5. Using a source of low pressure air, a pneumatic controller, and a pneumatic damper motor, connect the parts with pneumatic tubing to provide control of the damper according to the temperature of the air at the controller. First, operate the damper by changing the air pressure leading to the controller, and then operate it by adjusting the set point of the controller. Explain what caused the damper to move in each case.

6

Heating Unit Controls

Unit controls are controls installed on a heating or air-conditioning unit, usually by the manufacturer, for operating and regulating the operation of the various components and parts of the unit. Controls can be categorized in four groups: power controls, operating controls, safety controls, and in the case of combustion heating units, combustion safety controls. Unit controls are part of the total control system, and a designer must know how each operates so that it can be incorporated into the total control system design.

TYPES OF CONTROLS

Most heating unit controls are electromechanical controls; however, pneumatic and electronic controls may also be used. Unit controls usually provide the most fundamental control functions. They are used to control electrical power for motors and the control system, provide safe operation of all parts, and provide safe firing procedures for combustion furnaces. Some unit controls are common to many different types of furnaces, but others can be categorized by the type of heating system they are used on, such as gas fired, oil fired, or electric.

Controls Common to Many Heating Units

Controls that are common to many types of heating units are power controls, many operating controls, some safety controls, and combustion safety controls. *Power controls* provide for control of the electrical power used to operate most modern heating and air-conditioning equipment. *Operating controls* turn units, components, and parts of a system off and on, while *safety controls* provide for safe operation of a unit during normal operation or if a malfunction occurs. *Combustion safety controls* provide for safe firing of the fuels used in a heating unit.

Power controls. *Power controls* include disconnect switches and over-current protection devices (Figure 6-1). Disconnect switches are used to control the electrical current in a circuit. Disconnect switches are used at the unit, and

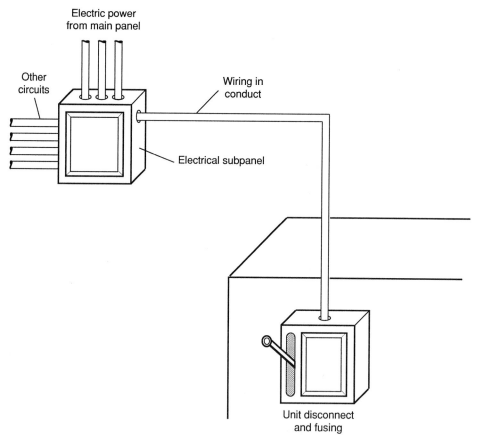

Figure 6-1 Power controls for a heating or air-conditioning unit are located in the electrical power lines to the unit.

Heating Unit Controls   Chap. 6

on branch electrical circuit feeding units, to turn the electrical power on for normal operation or to shut it off while servicing the unit.

Disconnect switches are either knife-blade or circuit-breaker types. A *knife-blade switch* has bladelike levers that are hinged at one end on terminals that are connected to the load side of an electrical circuit. The blades fit into contact terminals at the other end when they are closed. The contact terminals are connected to the line side of the circuit. To open the circuit, the switch blades are pulled open. To close the circuit, the blades are pushed closed. If the disconnect switch is part of a circuit breaker, it is a spring-loaded switch that can be opened and closed manually to open or close the circuit. The switch, with the circuit breaker, is wired into an electrical circuit the same as a knife-blade switch.

Overcurrent protection is provided by fusing or circuit breakers (Figure 6-2). If more electrical current flows in a circuit than elements of that circuit can carry, the fuses or circuit breakers open that circuit. A *fuse* is an electric

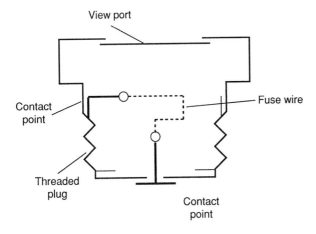

(a) Threaded fuse

Figure 6-2 Fuses or circuit breakers provide overcurrent protection to both unit and electrical lines.

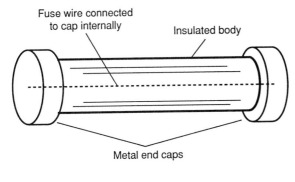

(b) Cartridge fuse

device in which a fuse wire of a specific size is located so that it is in series in an electric circuit. If the amperage in the circuit exceeds the fuse wire's current-carrying capacity, the wire melts and opens the circuit. In many cases circuit breakers are used instead of fuses for overcurrent protection. A circuit breaker is a spring-loaded switch that will open automatically if the amperage in the circuit exceeds a preset limit. It must be reset manually.

A *circuit breaker* is an automatic switch that opens if an overcurrent situation occurs. A circuit breaker can also be used for a disconnect switch. Disconnect switches and fuses are usually installed in each main circuit leading to a heating or air-conditioning unit in the main or a branch circuit panel. A disconnect switch is also located at the unit, so electrical power can be shut off during service on the unit.

Operating controls. *Operating controls* operate the parts of a heating unit, either singularly or along with other parts. They include the low-voltage transformer used to provide control power, thermostats or operators, solenoid or relay controls, fan controls, and any other control that operates components or parts of the unit. A *low-voltage transformer* is an electrical control that converts a higher voltage electrical signal to a lower-voltage one. The typical line voltage for a residential-size furnace is 120 volts. Commercial units may be 240 volts or even higher. A control transformer used on residential and many commercial furnaces will convert the line voltage to 24 volts for the control system. On some commercial units the control voltage is 120 volts or higher.

A transformer has two coils that are separated from each other electrically but are in very close proximity physically. The two coils are called *primary and secondary coils*. The two coils are wound around a metal armature so that their electromagnetic fields are interlocked when the primary coil is energized. An electric voltage is induced into the secondary coil by the electromagnetic field of the first coil. The primary coil has a larger number of turns of conductor around its coil than the secondary coil. The amount of voltage induced is proportional to the number of windings. In a low-voltage, step-down transformer such as that used for a control transformer, the ratio of windings is such that a step down of voltage provides 24 volts on the secondary side (Figure 6-3).

The thermostat used with most heating-only furnaces is a low-voltage thermostat (Figure 6-4). A thermostat is a temperature-actuated electrical switch. It is wired in the control circuit to the heating unit operating control so that it turns the furnace on if the ambient temperature drops below a predetermined temperature, and it turns the furnace off if the ambient temperature rises above a predetermined temperature.

A typical heating thermostat is built with a bimetallic sensing element. The bimetallic element is a strip or coil made of two metals with different coefficients of expansion. The metals are bonded together. If the temperature surrounding the thermostat rises, the element bends one way, and if the temperature falls, it bends the other way. The bimetallic element opens or closes a set of electrical contacts, either directly or through a mercury bulb switch.

Figure 6-3 A control transformer is used to reduce the voltage for the control system in a heating or air-conditioning system.

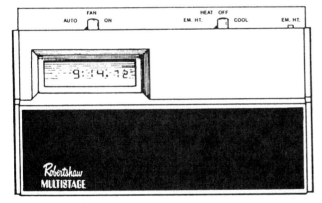

Figure 6-4 A thermostat is a basic operating control.

A mercury bulb switch is a glass bulb with a drop of mercury in it. Electrical terminals are embedded in the glass in one end of the bulb. If the bulb tips toward the end with the terminals in it, the mercury drop runs to that end and makes contact between the terminals. If the bulb tips the other way, the mercury runs away from the leads and opens the electrical circuit. The bulb is mounted on top of a helical bimetallic sensing element. The electric switch is wired in series in the control circuit of the heating unit, and the thermostat then starts and stops the unit in response to ambient temperature changes.

Types of Controls

Some heating units use a *remote-bulb thermostat,* in which the sensor is remotely located from the switch mechanism (Figure 6-5). Most remote-bulb thermostats are pressure operated. The bulb is a small container with a liquid such as a refrigerant in it. It is connected to the switch mechanism with a small-bore capillary tube. The tube is connected to a diaphragm or bellows element. As the pressure in the bulb changes in relation to temperature, the pressure in the tube moves the diaphragm or bellows element in or out. A push rod is connected to the back of the diaphragm or bellows, and this is connected to an electrical switch through a linkage assembly. The thermostat is calibrated to provide control through whatever temperature range is desired.

Solenoid controls use an electromagnetic solenoid coil as the operating element (Figure 6-6). Solenoid coils are used primarily to operate switches and valves. They are constructed of a hollow-core electromagnetic coil. A movable iron element is mounted so that it can move up and down within the hollow core. The iron element is attached to a switch or valve by mechanical linkage. If the coil is energized electrically, the iron element is drawn into the center of the core by electromagnetic action. This opens or closes the switch or valve as desired.

Relays are controls that make it possible to control a signal in one circuit with a signal from a separate circuit (Figure 6-7). Relays are used in all types of control systems. An electric relay is a control device that makes it possible to control one electrical circuit with a separate circuit. Electric relays are often used in systems in which the control voltage is different than the line voltage. An electric relay has a magnetic coil that pulls in an armature when it is energized. The armature has contacts on it that match other contacts in the control. As the armature moves back and forth when the coil is energized or deenergized, the contacts close or open. The control circuit of the unit is wired through the relay coil. The line circuit of the unit is wired through the relay contacts on the armature. If the coil is energized by the control circuit, the line circuit is either made or broken depending on the type of relay used.

Several types of electric relays are available (Figure 6-8). They are des-

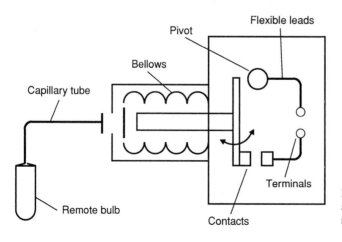

Figure 6-5 A remote-bulb thermostat allows control of temperature from a remote location.

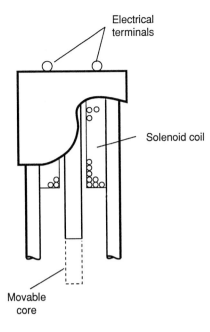

Electrical
terminals

Solenoid coil

Movable
core

Figure 6-6 A solenoid control is a magnetic device that is used for many operating functions.

Figure 6-7 A relay is an operating control that allows control of one electrical circuit by another.

ignated by the number of conductors in the circuit that they control, and the position of the contacts in the unenergized state. For instance, a relay that has a set of two contacts that are open when the relay is not energized is a two-pole, normally open (2P NO) relay. If the contacts are closed when the relay is not energized, it would be called a two-pole, normally closed (2P NC) relay.

Pneumatic relays allow air pressure in one line to control the air pressure in another line. For instance, controlled air pressure from a controller can be used to open or close a valve in a different air line.

An electronic relay makes it possible to use a low-voltage dc signal from a controller to open or close a circuit that controls line voltage for operating an actuator. A signal from a controller can turn a motor on or off.

Types of Controls

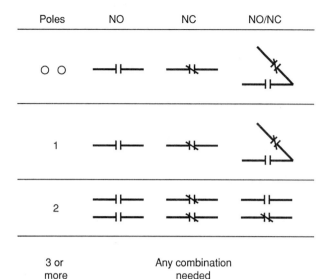

Poles	NO	NC	NO/NC
o o			
1			
2			

3 or more	Any combination needed

Figure 6-8 Relays are available in many different configurations.

Figure 6-9 A fan control is an operating control used to control the fan or blower in a system.

A fan control used on most heating units is a temperature-actuated electrical switch (Figure 6-9). A typical fan control has a sensor that extends into the airstream around the heat exchanger of a furnace. When the burner in the furnace turns on and the heat exchanger gets hot, the control turns the unit fan on. When the furnace goes off, the temperature around the fan control sensor cools off and the control turns the fan off. The sensor on a typical fan control is a bimetallic sensor similar to that used in a thermostat. As temperature changes occur, the bimetallic warps one way or the other and activates an electrical switch.

A secondary fan control may be needed on some horizontal and downflow

Heating Unit Controls   Chap. 6

furnaces. The secondary control, called a *sure-start fan control,* is used when the airflow through the furnace on startup may be the opposite of that desired because of the gravity flow of the air (Figure 6-10). This is usually the case in horizontal or downflow furnaces. A sure-start fan control has an electrical heater inside it that is wired in parallel with the heating unit actuator. When the thermostat calls for heat, the sure-start heater starts to heat up. The heater will close the fan control contacts within a certain length of time if the heat inside the heat exchanger does not.

A fan switch on a thermostat is often used for constant fan control (Figure 6-11). A thermostat with a fan switch built in to it is used, and the switch is wired in series with the coil on a relay. The contacts on the relay are wired in parallel with the contacts on the fan control on the unit. When the fan switch on the thermostat is turned "on," the blower motor is turned on through the relay. Motor controls and starting switches are controls that might be included with operating controls. In this book they are covered in the air-conditioner section in the discussion of compressor motors.

Safety controls.   *Safety controls* on a typical heating unit are installed to protect the parts, the entire unit, or surroundings from damage in case of a malfunction of the unit. Among the most commonly used safety controls are temperature limit controls and motor overloads. *Limit controls* are temperature-actuated electrical switches that open on temperature rise. They are normally installed in the airstream of the furnace at a point where the temperature of the air will indicate whether the unit is operating normally or not. If the air temperature increases beyond a set limit, the limit switch opens and shuts the

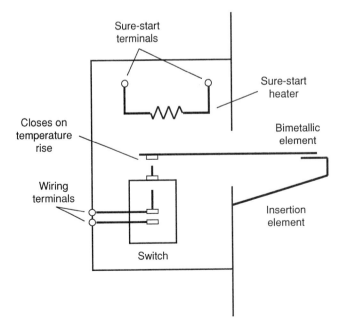

Figure 6-10   A sure-start fan control is used on downflow or horizontal-flow furnaces to ensure that the blower starts within a given period of time.

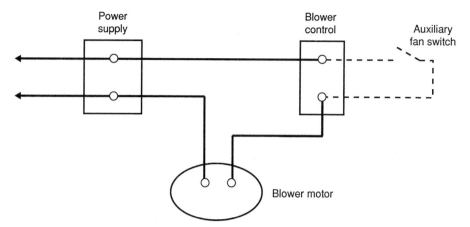

Figure 6-11 A switch may be added to the fan control circuitry to allow running a blower continuously.

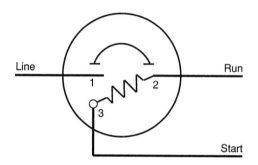

Figure 6-12 A motor overload will open the electrical circuits to the motor in case of malfunction.

unit down. A temperature rise would occur in the case of a malfunction such as the gas valve sticking open or the fan ceasing to function.

Secondary limits are extra limit controls installed on a heating unit to provide backup safety in case of a malfunction. If there is any reason to think that a situation could occur in which the primary limit would not sense an increase in temperature above that considered safe, a secondary limit would also be used. On applications requiring a downflow or horizontal furnace, if the blower motor failed or a blower belt broke, air could go through the unit in a direction in which the primary limit would not sense it. In this case a secondary limit on the other end of the unit would be used.

Motor overloads are temperature– or electric current–operated limits (Figure 6-12). They are used in the control or line wiring on an electric motor to protect the motor in case of a mechanical overload that would cause the temperature of the motor to get too hot, or in case of a short circuit or other problem that would cause too high a current draw. Either problem could damage the motor. A motor overload has a set of normally closed electrical contacts that open in case of temperature rise or current overload. Motor overloads are usually wired into the control circuit, so an overload will either turn off the motor directly, or will open the control circuit so that the motor is turned off indirectly in case of a malfunction.

There are three basic types of motor overloads: two wire, three wire, and four wire. They can be identified by the number of terminals. A *two-wire overload* is actuated by temperature alone. A bimetallic disk in the overload is connected by mechanical linkage to an electric contact. The contact presses against a second contact, and the two are wired into the control circuit for the unit. On an unusual temperature rise the bimetallic pulls the contacts apart, and the circuit is broken. A two-pole overload is usually mounted on the shell of a motor so that it is affected directly by motor temperature.

A *three-pole overload* is actuated by both temperature and by current overload. It has an electric heater inside its case that is wired in series with the motor terminals, and a bimetallic sensor with contacts in series with them. If an overcurrent condition or high-temperature situation occurs, the bimetallic sensor warps and the contacts are opened.

A *four-pole overload* contains an electric heater element and a bimetallic element with contacts. The heater circuit is separate from the circuit through the contacts. If either a current overload or high-temperature situation arises, the contacts open.

Controls Common to Gas-Fired Heating Units

Gas-fired heating units are those that burn a gas fuel. The basic unit controls on gas-fired units are similar to those used on equipment fired by other fuels. They are a low-voltage transformer, thermostat, limit controls, fan control, motor starters, and overloads. Some of the controls that are used specifically on gas-fired heating units are gas valves and particular types of combustion controls.

Gas valves. *Gas valves* used to control the flow of fuel on a gas furnace are controlled by the thermostat. Many different types of gas valves are used, but most of them have similar characteristics. The flow of gas through a gas valve is controlled by a valve that is controlled by either an electric solenoid coil or an electromagnetic coil. When the solenoid or coil is energized by the control circuit, the valve opens. When the coil is not energized, the valve is closed.

Most gas valves have a built-in pilot safety valve. This valve controls the flow of gas to the pilot flame and also provides safety shutoff in case the pilot light is not burning. A thermocouple is used to sense pilot flame, and this is connected to an electric coil on the pilot safety in the gas valve. If the pilot is not burning, the pilot gas valve remains closed and the main valve is locked closed. Only if the thermocouple senses heat can the pilot be lit and the main gas valve open. Many gas valves have a built-in pressure regulator. If there is not a pressure regulator in the valve, there will be one in the gas line ahead of it. There will also be a pilot safety control valve in the gas line if there is not one in the gas valve.

Combustion safety controls. *Combustion safety controls* used on a gas-fired heating unit ensure proper firing of the unit on a call for heat and protect against improper firing during a normal firing cycle. Several different types of combustion safety control systems are used. Some of the basic types are pilot safety, flame surveillance, and flame proving.

Combustion safety controls used are usually one of two types. They are electromechanical controls used with a thermocouple, or electronic controls used with some type of flame surveillance. The first is often used on models of heating units with standing pilot burner assemblies. The second is used on models of both standing pilot and intermittent pilot burners. Gas-fired heating units may use either atmospheric burners or power burners. An atmospheric burner is one in which all of the combustion air is supplied from ambient air, at atmospheric pressure, and part of the air mixes with the fuel prior to reaching the burner face. A power burner is one in which combustion air is supplied under pressure by a blower and combustion air mixes with the fuel at the burner face. There is a difference in the combustion safety controls used for the two types. Atmospheric burners may use either standing or intermittent pilots, so they may have either type of safety. Power burners usually use some type of flame surveillance or flame-proving safety.

Pilot safety combustion safety systems are used on furnaces that have standing pilots for firing the fuel on a heating cycle. These units generally have atmospheric burners. If a burner has a standing pilot, proving the pilot flame is sufficient to ensure firing of the gas when the gas valve opens on a call for heat. The most commonly used standing pilot safety is one in which a thermocouple is used to sense a pilot flame (Figure 6-13). A *thermocouple* is a control device that converts heat to electricity. It is made up of two wires of different

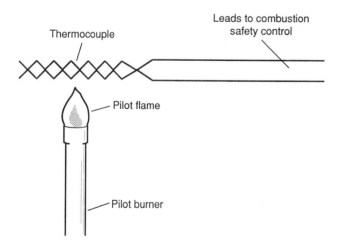

Leads to combustion
safety control

Thermocouple

Pilot flame

Pilot burner

Figure 6-13 A thermocouple provides an electrical signal to indicate that a pilot flame is lit.

types of metal twisted together. When the joined end, called a *couple*, is heated, electricity is produced. The thermocouple produces a low-voltage potential difference in the circuit formed by the extension of the wires. This low-voltage circuit is used to actuate a pilot valve in a gas valve. The pilot valve in the gas valve allows the main gas valve to open on a call for heat if the pilot light is heating the thermocouple. If the pilot light is not heating the thermocouple, the main valve in the gas valve will not open even if there is a call for heat.

Flame surveillance combustion safety control is achieved by using controls, usually electronic devices, that sense the establishment of a flame on a call for heat by actual surveillance. A typical example is a detector in which a photo-sensitive electric cell is used (Figure 6-14). A photosensitive electric cell is a light-sensitive cell in which the resistance to electric current flow varies de-pending on the intensity of light striking it. By placing such a cell where it can "see" the flame in a burner, it can be used to tell when a flame is established. An electric current is generated in a control center and circuited through the electric cell. If the cell senses the presence of a flame in the combustion chamber, the burner is allowed to come on. The resistance of the cell from the photo-electric cell is used to control the gas valve or oil burner on the heating unit. If a flame is not established on a call for heat, or if the flame goes out during normal firing, the cell indicates that there is no flame, and the valve or burner is shut off.

Flame proving is a combustion safety control that uses the flame from a gas burner as an electrical conductor. The particles of carbon in a flame make it an electrical conductor. A flame-proving combustion safety control has a conductor called a *flame rod* extend into the firebox, in the flame (Figure 6-15). This conductor is connected to one side of an electrical control circuit. The other side of the circuit is grounded to the burner. When a flame is es-tablished, an electrical circuit is completed through the flame. This circuit provides a signal to a safety control panel that causes the gas valve to close in case of a flame-out on a call for heat, or during normal operation.

Controls Common to Oil-Fired Heating Units

Oil-fired furnaces are heating units that burn heating oil as a fuel. Many of the unit controls used on oil-fired heating units are the same as those used on many other types of furnaces. Among these are transformers, thermostats, limit controls, motor starters, and overloads. Some controls that are common to most types of oil-fired furnaces are the combustion safety controls and some of the burner controls.

Combustion safety controls. *Combustion safety controls* are used to ensure that combustion occurs on a call for heat. In an oil burner, fuel and air are introduced into the firing chamber on a call for ignition. If a malfunction prevents firing, a dangerous condition is created as the combustible mix of oil and air fill the chamber. To prevent this combustion, safety controls shut the

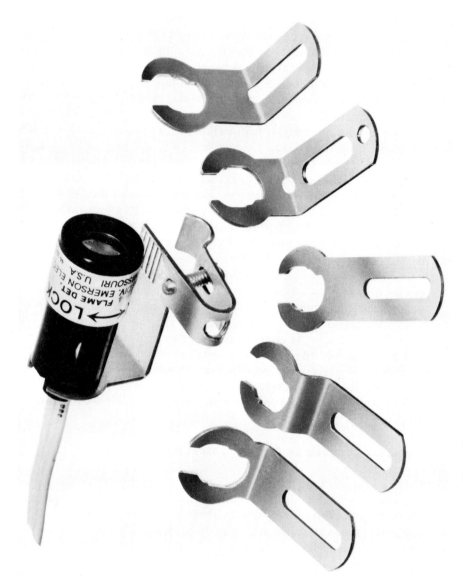

Figure 6-14 A cad cell is a photo-sensitive cell that is used to prove a flame.

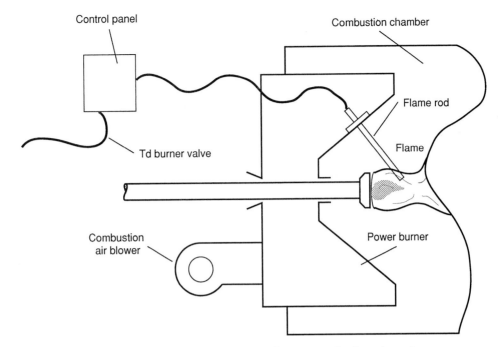

Figure 6-15 In a flame rod combustion safety system, the flame is used as an electrical conductor.

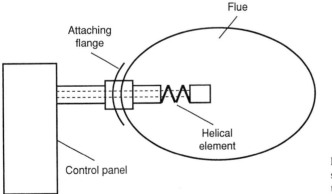

Figure 6-16 A stack switch is used to sense flue gas temperature for combustion proving.

unit down until the problem is identified and fixed. There are three basic types of combustion safety control systems used on oil-fired heating equipment: temperature sensing, flame surveillance, and flame sensing.

A typical temperature-sensing combustion safety control is called a *stack switch* (Figure 6-16). A stack switch combustion safety control is a control that senses flue gas temperature on a call for heat, and if the flue gas does not attain a temperature which indicates that the burner is firing, it shuts the burner down.

Types of Controls   **99**

If the flue gas temperature drops during a normal firing cycle, indicating a flame-out, the control will also shut the burner off.

A stack switch combustion safety control has a bimetallic element that is installed in the flue of the unit just where the flue gas leaves the unit. The bimetallic element is fixed on one end and the other end is connected to a movable rod. When the temperature of the flue gas goes up as the unit fires, the bimetallic element either rotates or bends, and the rod is moved. As the rod moves, it opens one set of contacts in a relay and closes another. The relay contacts are called cold contacts and hot contacts. The *cold contacts* start opening when the flue gas temperature rises, and the *hot contacts* start to close at the same time. The contacts are wired into the control circuit in such a way that one or the other of them must be made for the unit to fire. If the cold contacts open before the hot contacts close, the unit will shut off on safety.

Flame surveillance combustion safety control is achieved by using a control that senses the establishment of a flame on a call for heat by actual surveillance. A typical example is a detector in which a photosensitive electric cell is used. A photosensitive electric cell is a light-sensitive cell in which the resistance to an electric current flow is controlled by light striking the cell. The cell is located in an electric circuit originating in a combustion safety control panel. Variations in the electric current in the circuit are an indication of how much light is striking the cell. By placing such a cell where it can "see" the flame in a burner, it can be used to tell when a flame is established. The electric current from the photoelectric cell is used to control the fuel valve on a unit. If a flame is not established on a call for heat, or if the flame goes out during normal firing, the cell indicates that there is no flame, and the fuel valve is closed by the safety control.

Flame proving is a combustion safety control that uses the flame from an oil-fired burner as an electrical conductor. The particles of carbon in a flame make it an electrical conductor. A flame-proving combustion safety control has a conductor called a *flame rod* extending into the space where the flame occurs. This conductor is connected to one side of an electrical control circuit. The other side of the circuit is grounded to the burner. When a flame is established, an electrical circuit is completed through the flame. This circuit provides a signal to a safety control panel that causes the gas valve to close in case of a flame-out on a call for heat, or during normal operation.

Oil burner controls. *Oil burner controls* include the ignition transformer, combustion air fan control, and oil pump relay. Smaller oil-fired furnaces burn No. 2 burner oil, but larger units burn heavier oils. The controls on the different oil burning units are similar regardless of the size of unit or type of oil burned. Most oil-fired furnace burners are controlled by a control center called a *primary control* (Figure 6-17). The burner has several parts: oil pump, electric combustion, and combustion fan motor, which on a call for heat, are all turned on. These several parts are controlled at the same time by the primary control. The combustion safety control is also part of the primary control. The primary control contains a low-voltage transformer and a control

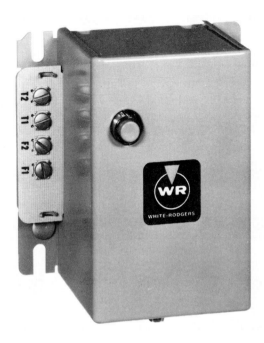

Figure 6-17 A primary control is part of a combustion safety control system.

relay to operate the oil burner, a lockout relay as part of the combustion safety control, a time-delay relay, and various contacts that function as combustion safety controls.

Controls Common to Electric Heating Units

Electric heating units are units that heat air with electrical resistance elements. The elements are heated when an electrical current passes through them, so control of the unit is basically control of the electrical current. Some of the controls on an electric heating unit are similar to those found on other types of heating units. They are the electrical power controls and the low-voltage transformer. Other controls are similar but modified in some way for electric heating use. These are the thermostat, control relay, stepped controller, limits, and fan switch.

A thermostat used with an electric heating unit is similar to that used for gas- or oil-fired units, with the exception of the heating anticipator (Figure 6-18). A heating anticipator heater has to be set to the amperage draw of the control circuit. An electric heating unit control circuit normally has very low amperage because the unit control relay is the only load in the circuit. To increase the amperage draw in the control circuit an electrical resistor is wired in the circuit, parallel to the relay coil. This increases the amperage enough so that a typical heating thermostat can be used. The resistor must be sized with

Types of Controls

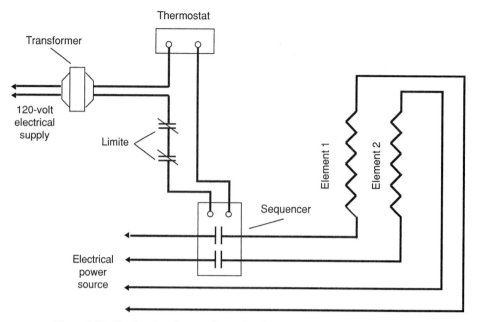

Figure 6-18 The heating elements in an electric furnace are usually controlled by a sequencer or stepped controller.

the proper ohm rating to increase the total amperage in the control circuit to somewhere between 0.10 and 1.0 A.

In some cases multistage thermostats may be used for controlling electric heating units, but staging is usually attained through the use of stepped controllers (Figure 6.19). A stepped controller is used on an electric heating unit to bring the heating elements on the line in steps instead of all at once. This ensures that the electrical load will be spaced out on a call for heat, and the electrical system will not be required to handle a large electric surge all at once. It also provides a form of temperature control similar to modulation. On a call for heat, since the heating elements are turned on one at a time, the heating load may be satisfied before all elements are turned on. In a particular application, if the heating load can be maintained by part of the elements being on, the others may not be turned on at all.

There are several types of stepped controllers used for electric heating elements. They range from mechanical to electric to electronic. *Mechanical stepped controllers* are usually constructed with several eccentric cams mounted on a shaft. The shaft is connected to an electric motor through a gearbox so that the rotational speed of the controller shaft is much less than that of the motor. The cams on the shaft are arranged so that the high points engage some sort of electrical switch mechanism. The cams are also arranged so that they engage the switches one at a time, in a particular sequence. The sequence can be selected as desired. Each switch is connected to one of the elements in the electric heating unit.

Heating Unit Controls   Chap. 6

Figure 6-19 An electrical sequencer is used as a stepped controller in many types of heating or air-conditioning systems.

Electric stepped controllers are normally thermally operated. They have a bimetallic element that is heated by an electric heating element when it is energized. There are a number of electrical switches arranged so that the bimetallic trips them as it warps. Electric controllers are also called multistage time-delay relays.

Electronic stepped controllers are control devices in which a time delay is introduced between various steps of a switching sequence. The time delay is a function of solid-state devices which provide output signals at a predetermined time interval, and the output signals are then used to actuate electrical switches that activate loads.

SUMMARY

Heating unit controls operate the individual parts on a heating unit. They are usually installed at the factory by the unit manufacturer on all smaller units. An exception to this is the thermostat or other operating controls that may need to be located in building spaces to be heated.

Many unit controls are common to all types of heating units. Among these are the power controls, some operating controls, and many safety controls. Variations are found mainly in combustion safety controls.

Power controls include disconnect switches and overcurrent protection. Operating controls include transformers, thermostats, relays, limit controls, fan controls, and different motor controls and overloads. Safety controls generally include temperature-limit controls and motor protection overloads. Combustion safety controls include thermocouples, stack switches, flame surveillance, and flame-proving devices.

QUESTIONS

6-1. Write a definition for unit controls as opposed to control systems.

6-2. Name three types of controls that are common to many kinds of heating and air-conditioning units.

6-3. Name two types of control devices that are used as power controls.

6-4. What feature of a circuit breaker makes it possible for it to be used to serve two purposes?

6-5. What is the main distinguishing feature of controls used primarily as operating controls?

6-6. *True or false:* A transformer has two coils that are connected to each other electrically.

6-7. *True or false:* There is no basic mechanical difference between a regular heating thermostat and a remote-bulb thermostat.

6-8. *True or false:* An electric relay is a control device in which one electrical circuit can be controlled by another.

6-9. *True or false:* An electrical fan control operates like a thermostat in that a set of electrical contacts are controlled by temperature.

6-10. Name the two most common types of limit controls used in electrical control systems.

6-11. What is the main purpose of motor overloads?

6-12. *True or false:* All motor overloads are actuated by temperature.

6-13. What is the purpose of the pilot safety feature on a gas valve?

6-14. Name three basic types of combustion safety controls used on gas-fired furnaces.

6-15. What type of pilot is usually used on furnaces that have a thermocouple as part of the combustion safety system?

6-16. *True or false:* A flame is a nonconductor electrically.

6-17. Name three basic types of combustion safety controls used on oil-fired furnaces.

6-18. *True or false:* A bimetallic element is part of a stack switch as used for a combustion safety control.

6-19. Name three functions of a primary control as used on an oil-fired furnace.

6-20. What is the main difference between a thermostat used on an electric furnace and one used on a gas-fired furnace?

6-21. What is the primary advantage of using a stepped controller to bring on the elements of an electric furnace one at a time?

APPLICATION EXERCISES

6-1. Write in the names of the parts of a remote-bulb thermostat on the accompanying diagram. Describe how the thermostat works, and explain why it is so useful in certain applications.

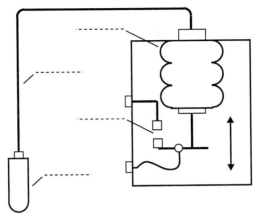

6-2. Name the relay shown in the accompanying diagram as to the number of poles and operation of the contracts. Draw a diagram showing how the relay can be used to lock-out one load while operating another load at the same time.

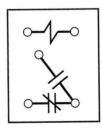

6-3. Draw in the wiring required to connect the relay in the accompanying diagram to allow the blower motor to operate at high speed off the fan control when the thermostat is calling for heat, but to run the motor continuously at low speed off the fan switch on the thermostat.

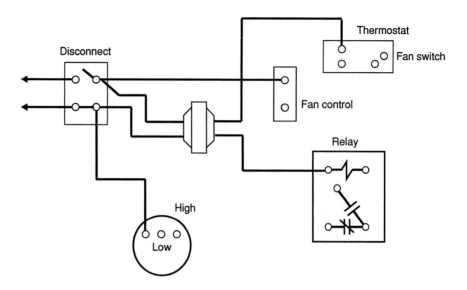

6-4. Name the major parts of a typical gas valve shown in the accompanying diagram by filling in the names in the blank spaces.

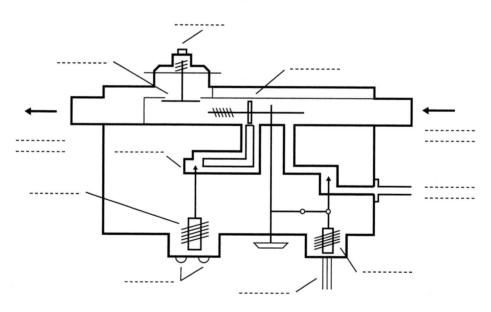

6-5. Name the major controls that operate the oil burner in the accompanying diagram.

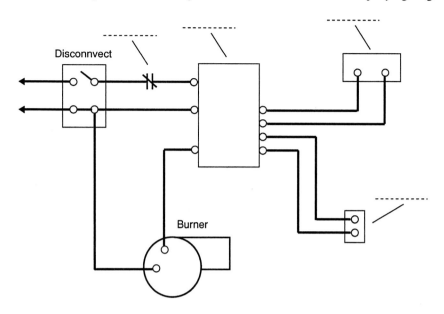

7

Air-Conditioning Unit Controls

Air-conditioning unit controls are controls installed on air-conditioning, humidification, or ventilation units. They are usually installed by the unit manufacturer. They are the controls that operate the various parts and components of the units. The term *air-conditioning equipment* can be used either to describe equipment that helps to condition air, or, as is more common in practice, equipment used to cool air. In this book the term will usually be applied as having the second meaning unless described otherwise.

Some air-conditioning equipment unit controls are similar to controls used on heating equipment. When that is the case in this chapter, the differences between the controls will be explained, but the basic control descriptions are given in Chapter 6. Unit controls for air-conditioning equipment can be categorized as power controls, operating controls, or safety controls.

TYPES OF AIR-CONDITIONING UNIT CONTROLS

Most controls used as unit controls on air-conditioning equipment are electro-mechanical, but pneumatic and electronic controls are also used. There are some differences in controls used on different types of equipment, such as that used for air cooling or water chilling, but the fundamental operation of the controls is usually similar. Such differences are covered in the various sections of this chapter.

Controls Common to Air-Conditioning, Humidity, and Ventilation Units

Unit controls that are common to many types of air-conditioning equipment are found in all three categories of controls. Electric power controls include disconnects and fusing. These are similar to those used in heating units, except that in many cases the controls must handle much larger electrical loads. Operating controls for air-conditioning units again are similar to those used in heating units. Thermostats for cooling equipment have to be designed so that they function on rising temperatures instead of falling ones. In most cases safety controls for air-conditioning units are different from heating unit controls.

Electric power controls. *Electric power controls* for air-conditioning equipment quite often have to carry larger electrical loads than do those described for heating units (Figure 7-1). Disconnects and fusing are similar but are sized for heavier electric current flow. The electrical current flow for any circuit is a function of the size of the electrical equipment serviced and is normally measured in amperes, the amount of current flowing in a circuit.

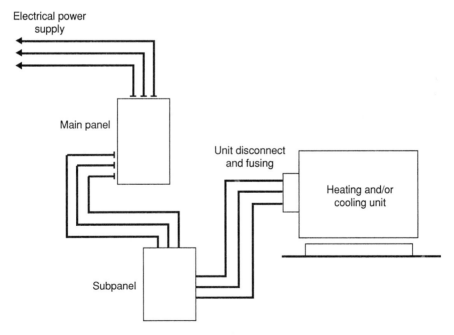

Figure 7-1 Power controls are controls that regulate the flow of electricity to a heating or air-conditioning unit.

Operating controls. *Operating controls* are used to turn various parts and components of a unit on and off. Those that are either different in design or are used differently from the ones used in heating units are control relays, cooling thermostats, and motor controls. Control relays are similar to those used for heating units, but in most cases the relays do not control the line-side current in a system directly, but only indirectly through another circuit (Figure 7-2). The secondary circuit controls a device called a *contactor* or *magnetic starter* that actually controls the on–off function of motors in the circuit.

A cooling thermostat is a control device that calls for cooling if the ambient temperature goes above a predetermined set point, and shuts off the cooling if the temperature goes below the set point. A typical cooling thermostat has a bimetallic element that warps as the temperature changes. As the element warps it opens or closes a set of electrical contacts. The contacts are wired in series with the cooling control relay in the control circuit.

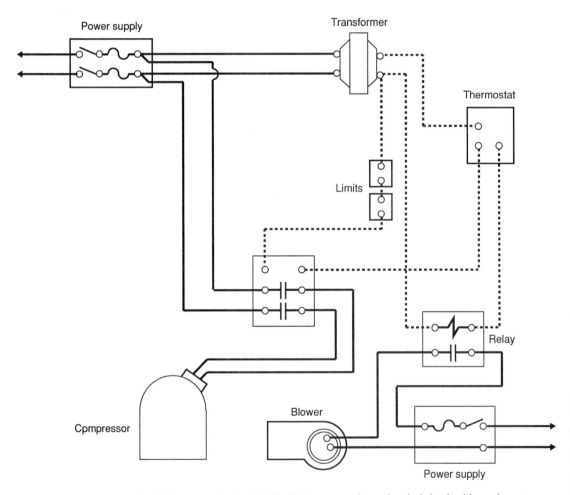

Figure 7-2 Relays are used where it is desirable to control one electrical circuit with another.

Air-Conditioning Unit Controls   Chap. 7

A cooling thermostat has an *anticipator* heater built into it to provide anticipation of the time when the ambient temperature will equal the set point (Figure 7-3). This anticipator is a small electrical heater wired in parallel with the control relay coil and thermostat contacts. The heater is energized when the thermostat is not calling for cooling. When the thermostat calls for cooling the heater is shunted out of the line and does not function. When a combination heating/cooling unit or system is involved, the thermostat is usually a combination heating/cooling thermostat. It has two bimetallic elements with separate switches, and functions for both heating and cooling.

A *contactor* is similar to a relay, but it has heavier contacts so that it can handle a larger electrical load (Figure 7-4). A contactor has a magnetic coil that pulls in a movable armature when it is energized. The armature has contacts on it that mate with fixed contacts on the frame of the control. The contacts are wired in series in the electrical circuit to a motor or other electrical load. The contactor coil is wired in series with the control circuit for the unit. When the control circuit is energized the armature is pulled in and the contacts complete the line circuit. The load is then energized.

A *magnetic starter* is similar to a contactor except that it has built-in overload protection for the circuit that it controls (Figure 7-5). A magnetic starter has electrical heaters in one or more of the lines in the load circuit. This heater opens a set of contacts in the control circuit if the current in the load circuit exceeds a set limit.

Several different types of magnetic starters are available. Those most often used with air-conditioning equipment are automatic-reset electromechanical

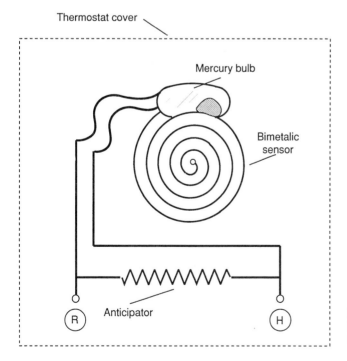

Thermostat cover

Mercury bulb

Bimetalic sensor

Anticipator

R

H

Figure 7-3 A cooling anticipator is a heater wired in parallel with the contacts in a cooling thermostat.

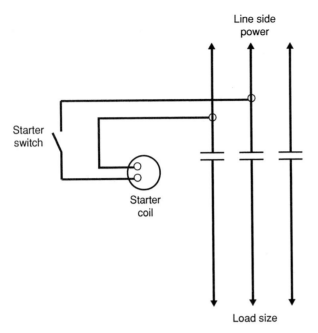

Figure 7-4 A contactor is used to control the electrical power to a motor.

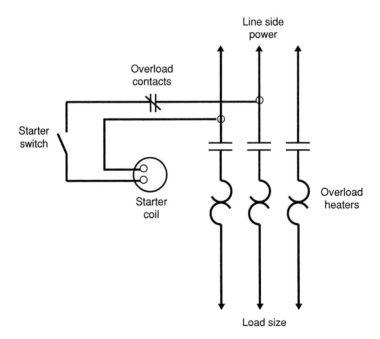

Figure 7-5 A magnetic starter has circuit overloads in it that act as current overload protection for circuits and motors.

devices. In an *automatic-reset starter* the contacts close automatically after the heater in the starter cools. In most cases an automatic-reset starter has some sort of signal indicating when the contacts are open so that a service person will know that it may come on again at any time. With larger motors, manual-reset starters are sometimes used. These are usually some type of relay requiring that an operator push a reset button to get the equipment back in operation after a safety shutdown.

Safety controls.   *Safety controls* shut a unit or system down in case of a malfunction. In some cases safety controls used on air conditioning are similar to those used on heating units, but most of them are unique to cooling and ventilation units. Some of the most commonly used are high- and low-pressure switches, oil pressure switches, and defrost controls for heat pumps.

High- and low-pressure switches are electrical control switches that sense refrigerant pressure in a refrigeration system and turn the unit off if the pressure goes too high or too low for safe operation (Figure 7-6). A *pressure switch* is a pressure-actuated electrical switch. It has a capillary tube that is connected to part of the refrigerant system. Typically, a high-pressure switch is connected to the hot-gas discharge line, and a low-pressure switch is connected to the suction line, or to the compressor crankcase if this is in the low-pressure side

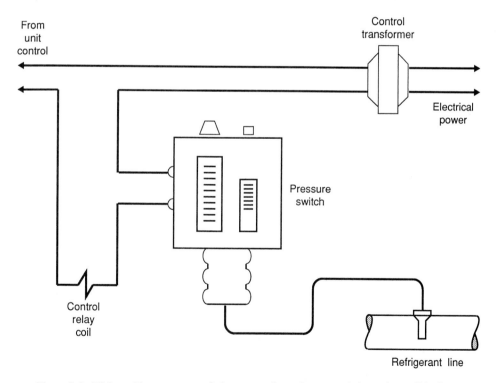

Figure 7-6   High- and low-pressure switches are used as safety controls in an air-conditioning system.

of the system. The capillary tube is connected to a diaphragm or bellows element on the control. The diaphragm or bellows is, in turn, attached through a mechanical linkage to a switch mechanism. The switch mechanism is adjusted by a set-point adjustment and a range adjustment. The set point is the pressure at which the switch opens, and the range is the difference between the opening and reclosing of the switch. The electrical contacts in the switch are wired in series in the control circuit for the unit. If the unit pressure goes above the set point on a high-pressure switch, the switch contacts open and the unit is shut off. When the unit is off, the pressure starts to drop. When it drops below the set-point pressure by the amount of the range setting, the contacts close and the unit comes back on if the thermostat is still calling for cooling. The low-pressure switch works in the same way except that the pressures sensed are low instead of high.

An *oil pressure switch* is a safety switch used on air-conditioning compressors large enough to have a detached oil pump for lubrication of the compressor bearings (Figure 7-7). Many large compressors (Figure 7-9), both reciprocating and reciprocal, have such lubrication systems. An oil pressure switch has a pressure-operated electric switch that is operated by oil pressure. If the oil pressure exceeds the set-point pressure, the electrical switch opens after an appropriate time delay. Most oil pressure switches have a time-delay feature built in so that the unit will not shut down on start up. The switch is wired in series in the control circuit to the compressor starter.

Defrost controls on an air-to-air or air-to-water heat pump are used to

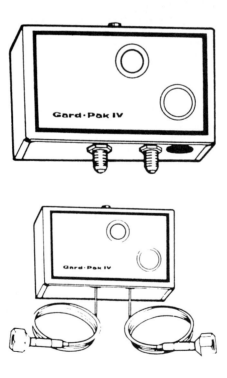

Figure 7-7 An oil pressure switch is used as a safety control when an air-conditioning unit has a positive lubrication system.

prevent the buildup of frost or ice on the outside coil during heating cycles (Figure 7-8). Because the outside coil is functioning as the evaporator coil, the temperature of the refrigerant is usually lower than the freezing temperature of water. Water vapor in the outside air will freeze on the coil during a normal heating cycle. To remove this, a defrost control is used that shuts off the outside fan, puts the unit in a cooling mode, and turns on the auxiliary heat source.

The defrost control is usually designed with a clock timer that initiates a defrost action in regular time cycles. The timer can be set so that the action is initiated every 20 minutes, 40 minutes, or 1 hour. When the action is initiated, a temperature sensor on the outdoor coil is checked to determine if a defrost cycle is needed. If it is, the defrost control takes over. If the sensor shows that the temperature of the coil is high enough to indicate that no defrost is needed, the defrost is terminated and the clock timer continues to the next time interval.

When a defrost cycle is in effect, the temperature sensor indicates when the cycle can be terminated. As soon as the temperature of the coil goes high enough to indicate that there is no frost or ice on the coil, the defrost cycle is terminated. Most defrost controls are combinations of electronic and electro-mechanical controls.

Controls Specific to Air-Conditioning Units

Air-conditioning units come in different types depending on the method used for attaining a low temperature and on the medium used for cooling: air or water. There are two different general ways to attain low temperatures: one by

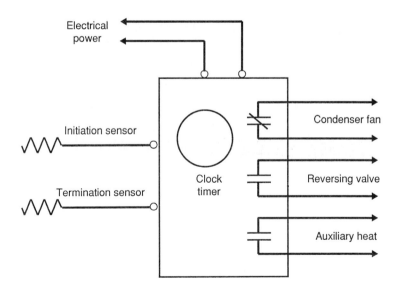

Figure 7-8 Defrost controls are used on a heat pump to make sure that ice does not build up on the outdoor coil.

use of a mechanical compression system, the other by use of an absorption system (Figure 7-9).

Mechanical compression units. *Mechanical compression refrigeration* is a method of refrigeration used to attain low temperatures in which a gas called a *refrigerant* is compressed by mechanical means, and then liquefied and cooled by use of a condenser. The high-pressure, relatively cool refrigerant liquid is then allowed to expand rapidly into an evaporator coil. As the refrigerant expands, its temperature drops and the refrigerant boils off into a vapor. The rapidly expanding refrigerant picks up heat from the coil. Air or water passes over the coil and is cooled in the process. The air or water is then used as a cooling medium.

Controls that are used specifically on mechanical compression air-conditioning units differ somewhat depending on whether the unit is air cooled or water cooled and whether the unit uses air or water as a cooling medium. Air-to-air system controls typically used on mechanical compression air-conditioning units include low ambient controls and an off-cycle timer.

Low ambient controls allow the air-conditioning system to run when the outside temperature is low enough to cause the refrigerant temperature to drop low enough to cause frosting or icing on the evaporator coil (Figure 7-10). One type of low ambient control system uses a low-temperature thermostat and relay to shut off the condenser fan if the outdoor temperature gets too low. The relay is a normally closed relay with the contacts wired in series in the line circuit to the outdoor fan motor. Another type of low ambient control uses dampers to control the flow of air across the condenser coil. This control system is made up of a low temperature thermostat and damper operators. The dampers are closed when the outdoor temperature gets too low.

An *off-cycle timer* is a timer device used to prevent a compressor from short-cycling. If a compressor motor is allowed to turn on and off frequently, the motor can overheat and be damaged. A timer is wired into the compressor electrical circuit in such a way that when the compressor shuts off after a normal cooling cycle, it cannot come back on for at least 5 minutes. The controls

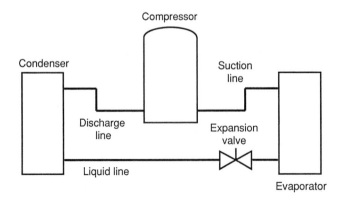

Figure 7-9 The basic components of an air conditioner are the compressor, condenser, evaporator, expansion device, and refrigerant lines.

Air-Conditioning Unit Controls   Chap. 7

From control
transformer

Remote-bulb
thermostat

NC relay

Bulb in
outdoor air

Figure 7-10 Low ambient controls are
used on an air conditioner to prevent
the condenser blower from running
when the outdoor temperature is too
low.

Outdoor fan
circuit

include a timer that operates on a 5-minute cycle each time the unit shuts off,
and a normally closed relay that opens during the off time. The relay is wired
into the control circuit to the compressor relay or contactor.

Water-to-air system controls specific to mechanical compression equipment are related primarily to the water side of the system (Figure 7-11). A
water-to-air air-conditioning unit is one that uses water as a condensing medium
and air as a cooling medium. The controls required are related primarily to
water flow through the condenser.

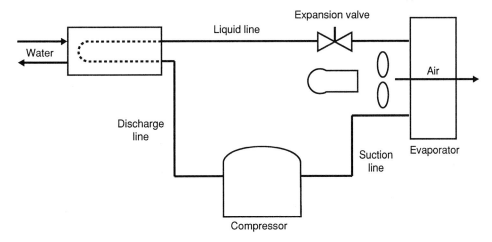

Expansion valve

Liquid line

Water

Air

Discharge
line

Suction
line

Evaporator

Compressor

Figure 7-11 A water-to-air conditioner uses water as a condensing medium.

Types of Air-Conditioning Unit Controls   **117**

The first consideration is for a constant supply of water. This is usually provided from a cooling tower system or from wells drilled for the purpose. Assuming that the supply of water is assured, control of the water is the next consideration. The flow of water through the condenser coil must be controlled to provide the proper amount of heat removal for the cooling effect desired. Since the cooling effect changes as the load on the building changes, the water flow must be adjusted to match that load.

An automatic water valve is used to control the flow of water through a water-to-air condenser coil (Figure 7-12). Automatic water valves are usually remote bulb pressure or temperature-actuated mechanical valves. A pressure-actuated water valve has a capillary tube tapped into the liquid refrigerant line leaving the condenser. The capillary tube is connected by a mechanical linkage to a bellows element on a water valve. The valve is located on the water line

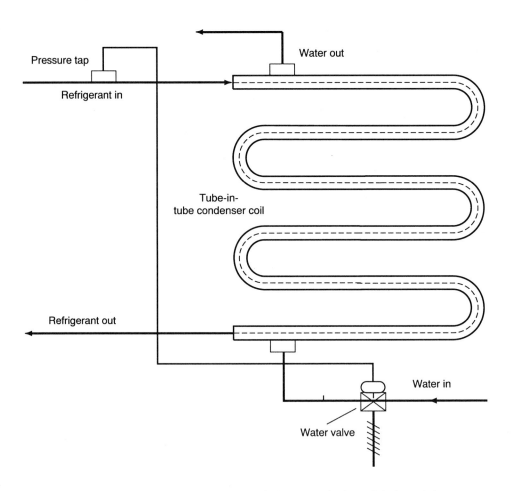

Figure 7-12 The water flow valve on a typical water-to-air air-conditioning unit is controlled by refrigerant pressure.

leading to the condenser coil. It is the pressure of the refrigerant in the liquid line that controls the flow of water. If the refrigerant pressure starts to drop, indicating that too much heat is being removed in the condenser coil, the water flow rate decreases. The water then removes less heat, and the refrigerant pressure increases. If a temperature-actuated valve is used, the capillary tube has a bulb on the end that is filled with refrigerant. Changes in temperature cause the pressure in the bulb and capillary tube to go up or down. The power head on the valve is similar to that used with a pressure-actuated valve. The water flow is controlled by the temperature of the refrigerant in the liquid line.

Air-to-water air conditioning controls are different from typical air-conditioning controls in that the temperature of the water leaving the unit is controlled. An air-conditioning unit in which water is cooled is normally called a *chiller*. Since the water temperature leaving the unit is controlled, the main operating control is called an *aquastat* (Figure 7-13). An aquastat is similar to a thermostat but the sensor senses water temperature rather than air temperature. An aquastat is usually a remote-bulb unit. The bulb, which is immersed in the water, contains a refrigerant. The bulb is attached to a diaphragm or bellows element. As the temperature of the water around the bulb changes, the pressure in the bulb, and consequently in the capillary tube, also changes. The pressure changes are transmitted to the diaphragm or bellows element, which, in turn, is connected to an electrical switch by mechanical linkage. The electrical switch is wired in series in the control circuit for the chiller compressor motor.

The circulation of water in a water-cooled air-conditioning system is usually

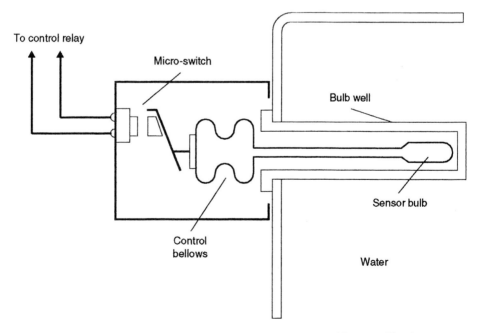

Figure 7-13 An aquastat is a temperature-actuated switch used for controlling the temperature of the water in a boiler or chiller.

Types of Air-Conditioning Unit Controls

controlled by thermostats in various spaces in the building. The thermostats turn circulating pumps on and off in each zone, or controlled space, to maintain the set-point temperature in the space. Water-to-water air-conditioning controls are basically similar to those described for the other three types of systems above (Figure 7-14).

Absorption systems. The second type of air-conditioning system is an *absorption system.* The refrigeration system in an absorption air-conditioning unit works on the same principle of pressure and temperature difference on which a mechanical compression unit works. The pressure and temperature differences of the refrigerant in the unit are attained through the use of a chemical that absorbs the refrigerant on the high-pressure side of the system, and separates from it on the low-pressure side of the system. The controls in absorption air-conditioning units are primarily solenoids, relays, and various pump controls. They are similar to those already described, and because of differences in different units, no attempt will be made here to cover specific controls used in the units.

Controls Specific to Humidification Units

The control of humidity, the amount of moisture in the air, in a building is of prime importance in keeping the building comfortable. Too much humidity makes air feel clammy and uncomfortable, and too little moisture makes the

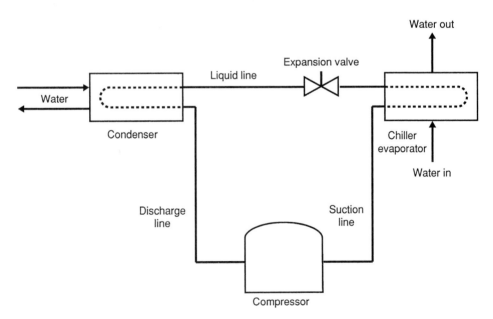

Figure 7-14 A water-to-water air-conditioning system uses water for both a condensor medium and an evaporator medium.

air feel dry and cold. Equipment is available either to add moisture to the air or to remove moisture from the air. Controls must be used to operate this equipment to provide the proper level of control.

Humidification controls. *Humidification,* which is the addition of moisture to the air, is usually achieved by adding water vapor in some form. Water vapor may be added to the air by evaporation or more directly by atomization. The equipment used may be part of the central heating, ventilating and air-conditioning, (HVAC) system, or it may be in the form of small package units. The systems used may be categorized as follows: water curtain, steam, or package units.

A *water-curtain humidification system* is one in which a virtual curtain of water is maintained in the airstream of a system downstream from a heating coil or source (Figure 7-15). As the supply air for the spaces to be humidified passes through the curtain, water is evaporated and the air is humidified. The flow of water is controlled by a control valve, which, in turn, is controlled by a humidistat. The control valve is a typical liquid flow control valve powered by whatever type of control system is used. The humidistat is a control device that has a sensor that senses the humidity level of the ambient air. The sensor may be a natural product such as human hair, or wood, or it may be a manufactured product such as nylon. If an electronic control system is used, the sensor may be a chemical or metal alloy that senses humidity levels. The sensor operates a switch or other operating device to provide a control signal to the system actuator.

A *steam humidification system* has a steam injector nozzle inserted in the air supply duct in the distribution system for a building (Figure 7-16). Steam flow to the injector is controlled by a steam valve, and the valve is controlled by a humidistat, as described above.

Package humidifiers are self-contained units that introduce water vapor

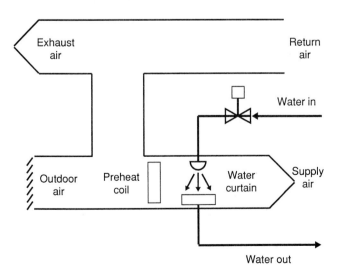

Figure 7-15 A water-curtain humidifcation system uses water as a direct source of humidity.

Types of Air-Conditioning Unit Controls **121**

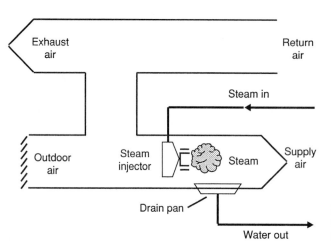

Figure 7-16 A steam humidification system literally injects steam into the air stream going into a building.

into the air in some form or other (Figure 7-17). Some are evaporation units in which water is either evaporated directly by a heating unit, or soaks a sponge or filter and is evaporated by air passing through the filter. Other package humidifiers introduce water vapor into the air by atomization. A small stream of water is directed onto a spinning wheel and is atomized by the contact. The atomized particles of water are introduced into a stream of air that is directed into the space being humidified. Whatever type of humidifier is used, the flow of water is controlled by a control valve on the water line and a humidistat that senses humidity in the air. The humidistat either opens the valve or closes it for humidity control in the space.

Dehumidification controls. *Dehumidification controls* are used to control equipment that removes moisture from the air to maintain a comfortable atmosphere in a building. The most common way to remove moisture is by use of an air-conditioning system. The air is naturally dehumidified as it passes over the cold evaporator coil in an air-conditioning unit. Other ways of dehumidifying are with chemical filters that absorb moisture or by molecular filters that adsorb moisture.

Air-conditioning controls for dehumidifying include a humidistat to control the operation of the unit instead of a thermostat, and an operating relay. Controls for chemical or molecular filters are a humidistat and relay to operate the blower for the unit, and in some cases controls for the regeneration of the absorbant or adsorbant elements.

Controls Specific to Ventilation Units

Ventilation of most buildings can be taken care of most satisfactorily by the introduction of outside air (Figure 7-18). Outside air is introduced into the building distribution system, where it is mixed with return air. Control begins

Figure 7-17 Most residential humidifiers are self-contained units.

at the point at which the air is brought in from outside. This is generally through a set of dampers. The dampers themselves are controlled manually for the simplest type of system. More often, automatic control is provided, where a mixed-air controller and outdoor-air controller are used.

In an automatically controlled damper package, return air from the building and outdoor air are both ducted to a mixed-air plenum. An exhaust-air damper and fan may or may not be used in the return-air duct ahead of the return-air damper. Return-air control is achieved by a set of dampers in the return-air duct just ahead of the mixed-air plenum of the damper package. Outside-air control is by a set of dampers in the outdoor air duct, just ahead of the mixed-air plenum (Figure 7-19). Both the return-air and outdoor-air dampers are controlled by damper actuators controlled by a mixed-air controller located in the mixed-air plenum of the damper package. The mixed-air con-

Types of Air-Conditioning Unit Controls

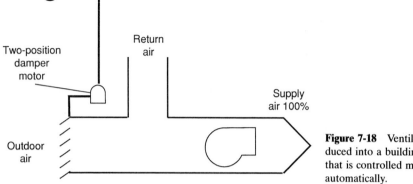

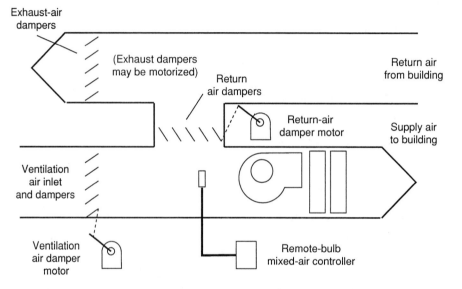

Figure 7-18 Ventilation air is introduced into a building through a damper that is controlled manually or automatically.

Figure 7-19 An economizer damper system is controlled automatically by a mixed-air control to free cooling when the outdoor temperature is low enough to cool a building.

troller is usually a remote-bulb controller that senses the mixed-air temperature and adjusts the outdoor- and return-air dampers to maintain the mixed-air temperature desired. In a typical control application, the mixed-air temperature setting is the same temperature as the cooling-cycle temperature desired for the air entering the building spaces.

An outdoor temperature control is used to override the mixed-air controller if the outdoor air gets warmer than the return-air temperature. On override, the outdoor damper returns to a minimum position. The outdoor temperature control is usually a remote-bulb controller. The damper motors for the system can be two-position, or modulating. The controller and motors are matched for proper performance.

SUMMARY

Controls used on air-conditioning units are designed to operate the equipment automatically and safely to produce a cooling effect on the air in the building in which it is used. Three basic types of controls are used: power controls, operating controls, and safety controls. Some of these controls are common to all types of air-conditioning units, whereas others are specific to a particular type of unit.

Controls functions are generally related directly to the type of air-conditioning units on which they are used. There are four types of units: air-to-air, air-to-water, water-to-water, and water-to-air. Most of the differences in controls relates to the medium being controlled.

Humidity and ventilation controls are included in our discussion of air-conditioning controls because of their use to control air conditions directly.

QUESTIONS

7-1. Define the term *air conditioning* as used in this chapter.

7-2. Name three categories of controls found in air-conditioning units.

7-3. What is one difference between the power controls used in air-conditioning units and those used in heating units.

7-4. Name the two primary differences between cooling and heating thermostats.

7-5. What main feature does a magnetic starter have that a contactor does not have?

7-6. *True or false:* A high-pressure switch in an air-conditioning unit shuts the unit down if the pressure does not go high enough.

7-7. Why are defrost controls necessary on a heat pump unit?

7-8. Name two types of controls found on air-to-air air-conditioning units.

7-9. *True or false:* Low ambient controls allow an air-conditioning unit to run when the outdoor condensing temperature is lower than desirable.

7-10. *True or false:* An off-cycle timer keeps an air-conditioning unit from coming on except when it is cold outside.

7-11. On a water-cooled air-conditioning unit, what is the name of the valve used to control water flow through the condenser?

7-12. A unit that adds water vapor to the air to maintain a specific humidity level in a building is called a _____ (humidifier/dehumidifer).

7-13. Name three different types of humidification systems typically used.

7-14. What is the most common type of dehumidifier used?

7-15. Explain what a mixed-air plenum is in a ventilation system.

7-16. What purpose does an outdoor temperature control serve in a ventilation system?

APPLICATION EXERCISES

7-1. Identify the main parts of an electric cooling-only thermostat by writing in the part name on the attached diagram. Describe the function of each part, and explain what its function is in controlling an air-conditioning unit.

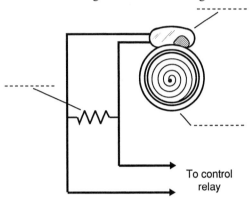

7-2. Complete the wiring on the attached diagram showing how a magnetic starter would be wired to a motor in an air-conditioning unit.

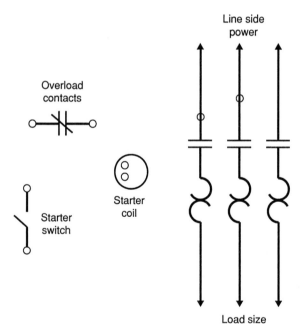

7-3. Mark the location with arrows for a cutoff pressure of 127 psf and differential of 15 psf on the accompanying facsimile of a low pressure switch used on an air-conditioning unit. Identify each location as to whether it represents cutoff pressure or differential. Explain how the differential pressure relates to the cutoff pressure.

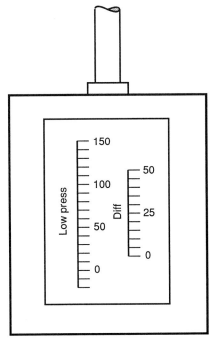

7-4. Fill in the spaces on the attached diagram of a heat pump defrost control system to identify the parts of the heat pump that are turned off or on in response to signals from the defrost control system.

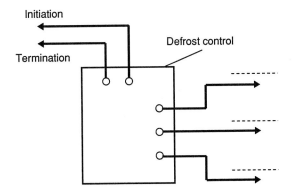

7-5. What is actually controlling the flow of water through the water valve, in this application on a water cooled air-conditioning unit?

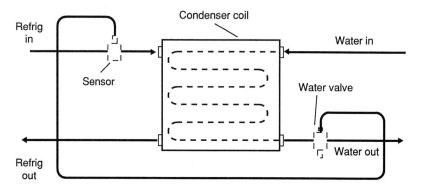

7-6. Name the four main parts of the absorption system shown in the accompanying diagram and describe what happens to the refrigerant/absorbant mix in each part.

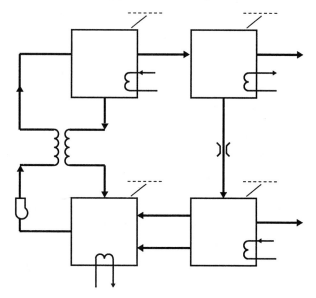

8

Manual Controls

Manual controls are controls that have to be set manually. On some applications they are used to operate or regulate the operation of heating or air-conditioning equipment. Few manual controls are used on modern heating and air-conditioning systems, and most of them are used in the electrical power controls. Manual controls are used as motor controls, as manual reset on some combustion safety controls, and as seasonal changeover controls.

ELECTRICAL POWER CONTROLS

All heating and air-conditioning systems that use electricity as an energy source have manually operated disconnects of some sort in the electrical supply lines to the equipment. Disconnects are provided at the main electrical supply panel, at any branch supply point, and at the units themselves. These switches are mechanical devices and are manually operated. The disconnects are normally used only when the power supply is turned off to the unit for maintenance or service.

Disconnect Switches

Disconnect switches are usually one of two types: blade switches or part of a control called a circuit breaker.

Blade switches. A *blade switch* has a blade that is an electrical conductor for each lead in the electrical circuit. Each blade pivots out from the terminal at one end. This end is connected to the load side of the circuit, the side that is connected to the equipment in the system. The other end of the blade fits into a tight-fitting socket when it is closed. This socket part of the electric circuit is connected to the line side of the system, the electrical supply side. To open the electrical circuit the blades are pulled open, away from the sockets they fit into. To close the circuit the blades are closed, or pushed into the sockets. The blades on a blade disconnect switch are always controlled by a handle that is insulated from the electrical circuit (Figure 8-1).

Blade disconnects are usually mounted in a protective metal box. The electrical connections are made inside the box. Some disconnect switches can be opened and closed by use of a handle on the outside of the switch box. Other disconnect switches have the operating handle inside the box and require that the box be opened to operate the switch.

Circuit breakers. A *circuit breaker* is a control that provides current overload protection for an electrical circuit while functioning as a disconnect switch (Figure 8-2). Overload protection is provided by an internal spring-loaded switch that opens if the electrical current going through it exceeds the current rating for the control. If the switch opens in an overcurrent situation, it can be reset and closed again after the situation is corrected. A separate circuit breaker is used on each leg of an electrical circuit. The internal switch in a circuit breaker can be opened manually so that the breaker functions as a disconnect for the electrical lead on which it is used.

When a circuit breaker is used as a disconnect switch for an electrical circuit to a heating or air-conditioning unit, one breaker is wired into each of the power leads to the unit. The electrical current to the unit must pass through the breakers. Opening any of the breaker switches opens the circuit to the unit, and turns it off.

Figure 8-1 A disconnect switch is a power control used to disconnect a unit from the power source when necessary.

Figure 8-2 A circuit breaker combines the functions of disconnect switch and overcurrent protection.

BR120 BR230 BR340 GFCB120 BJ3200

ELECTRICAL MOTOR CONTROL

When used in connection with heating and air-conditioning systems, the term *mechanical electrical motor control* usually refers to control of fan motors on air circulation systems or control of compressor or condenser motors. Motors may either be turned on and off or may be controlled so that their speed varies. Motor control is achieved by the use of a manual switch, or through a magnetic contactor or magnetic starter actuated by a manual switch. Either type of control achieves the same results, but the wiring and devices used vary somewhat.

Direct Manual Control

Direct manual control is control of a piece of equipment directly by use of a switch in the electrical lines to the equipment or in the control circuits to the equipment (Figure 8-3). This switch must be of the snap-acting type to prevent electrical arcing during switching. Direct manual on–off motor control requires the use of a manual starter switch located in the line side of the electrical circuitry to the motor or in the control circuit to the motor. This switch is normally of the blade type, similar to a disconnect switch. Since no inherent motor protection devices are built into a disconnect switch, a manual motor starter would normally be used on a motor with internal overloads only. Internal motor overloads are used to protect the motor and electrical circuit in case of a current overload.

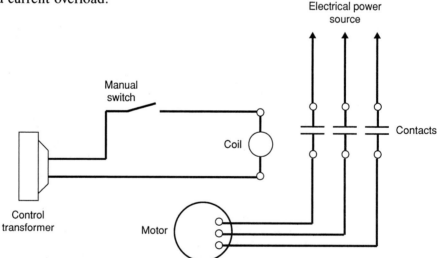

Figure 8-3 The operation of most electric motors is by a contactor or starter energized through a control circuit.

Indirect Manual Control

Indirect manual control of a motor is through a relay or starter device controlled by a manual switch. *Magnetic contactors* are the most commonly used controllers for starting and stopping electric motors, through use of a remote electric control switch. Magnetic contactors have normally open (NO) contacts that close when energized by a control circuit. The control circuit may be low-voltage or line-voltage electricity.

Magnetic contactors. In a *magnetic contactor* a set of contacts are wired in series between the electrical power line and the load, or motor. The contacts are opened and closed as a magnetic coil is energized or deenergized. The coil is wired in series in an electrical circuit with a power source and an electrical switch. If the electrical switch is closed, the magnetic coil is energized and it pulls in the contacts so that the motor operates (Figure 8-4). If the electrical switch is opened, the magnetic coil is deenergized. This allows the contacts to open and the motor stops. A magnetic contactor is used with motors that have inherent overload control. Since the contactor has no current protection device, the motor overloads provide that protection for a motor and its electrical circuits.

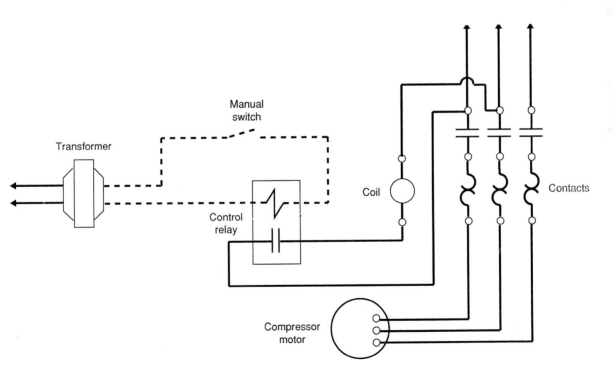

Figure 8-4 A magnetic contactor with a line-voltage coil may be energized by low voltage by use of a relay.

Magnetic starters. A *magnetic starter* is similar to a magnetic contactor except that a magnetic starter has built-in overload protection for a motor and the electrical circuits connected to it (Figure 8-5). The magnetic starter has a set of normally closed (NC) contacts in series with the electromagnetic coil that actuates the line contacts (Figure 8-6). The contacts in the control circuit are actuated either by a current overload in the line circuit or by a temperature increase in the motor windings. Either occurrence is an indication of a problem with the motor or in the motor circuits.

Directional Control

Control of the direction of the rotation of some motors is by arrangement of the internal connections of the windings. Connecting wires coming from the field coils in the motor in one configuration results in clockwise (CW) rotation, while connecting them in the opposite configuration results in counterclockwise rotation (CCW) (Figure 8-7). To reverse the wiring connections a manual switch is used. When the switch is set in one direction, the motor runs one way, and when the switch is moved the other way, the motor reverses.

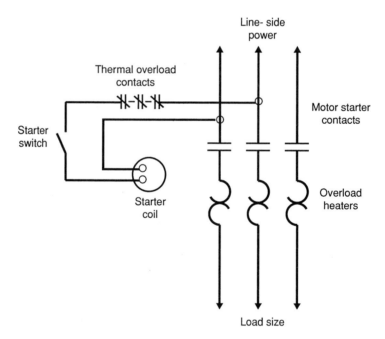

Figure 8-5 A magnetic starter may be controlled by a line-voltage switch.

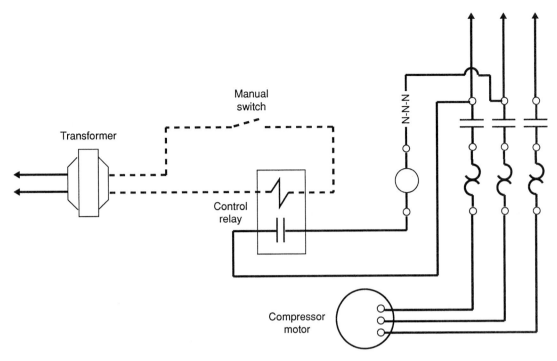

Figure 8-6 Low-voltage controls can easily be interfaced with line-voltage controls by use of appropriate relays.

Speed Control

Speed control of electric motors can be achieved by several different means. The means chosen for a given application usually depends on the type of motor used.

Voltage regulators. Speed control in some fractional horsepower motors can be achieved by using a *voltage regulator*. The regulator is usually called a *speed controller*. The controller is connected to the electrical lines running to the motor and can be set for different speeds. This type of speed control does reduce the efficiency of the motor and should be used only on certain types of smaller single-phase electrical motors. This type of speed control was often used on direct-drive blower applications when these blowers were first used.

Multitap windings. Some multispeed motors have *multitap windings* (Figure 8-8). They are manufactured with the running coils separated into several segments. Wire leads run from each segment to a terminal connector. In effect, the different segments of the windings change the number of running coils in the field of the motor. Power applied to each segment of the winding causes the motor to run at a specific speed, each differing from the speed of

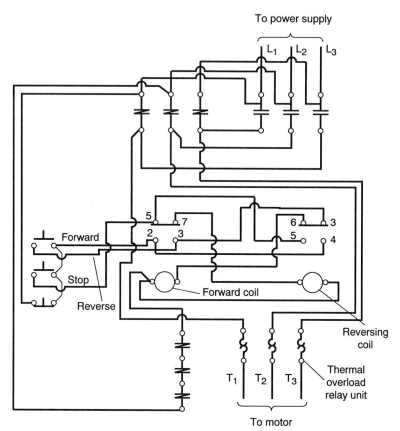

To power supply

L_1 L_2 L_3

5 7
2 3

6 3
5 4

Forward

Stop

Reverse

Forward coil

Reversing coil

Thermal overload relay unit

T_1 T_2 T_3

To motor

Figure 8-7 Bidirectional control of an electric motor requires, in effect, two starters or contactors and a reversing switch.

al other segments. By connecting power to any one segment, a specific speed of rotation is achieved. In application the motor winding connections are made so that the motor will run at the speed desired. Multiwinding motors are normally available in from two to five speeds. They are typically used in direct-drive blowers in furnaces and blower coil units.

Electronic speed controllers. Speed control for larger electric motors, three phase over 1 hp, is best achieved by use of an *electronic speed controller*. The electronic speed controller automatically adjusts the speed of the motor to match the load imposed on it. In many applications, such as variable-volume air systems, it is important to match the cubic-feet-per-minute (cfm) output of a fan to the air requirements of the system. In the typical fan used for circulated air systems, the cfm output is a function of the blower wheel speed. The best way to adjust the blower speed is by changing the motor speed. An electronic speed controller will save a considerable amount of electrical energy while it controls the speed of a motor effectively.

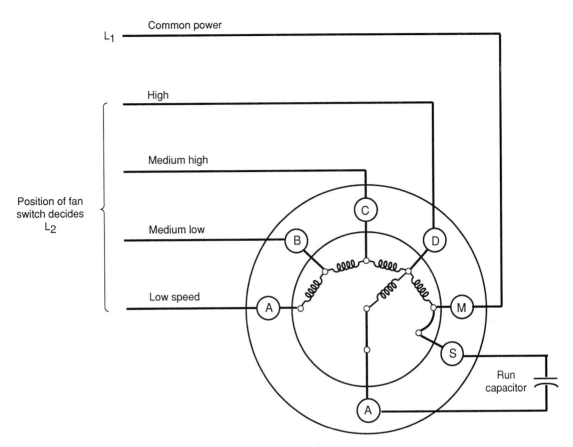

Figure 8-8 Multiwind motor connection.

Combustion Safety Reset

Most combustion safety control systems require manual reset in case of a flame-out. For a combustion furnace, a flame-out occurs any time the burner does not fire on a call for heat. This condition usually indicates a dangerous condition. If a combustion safety control resets itself, fuel and air could flood into the furnace combustion chamber and create an explosion hazard. To prevent this, the combustion control reset is manual. In case of a safety lockout, the unit should be checked before the reset is pushed, and the cause of the lockout should be corrected.

A combustion safety reset is usually incorporated into the combustion safety controls used on a furnace. The relay contacts of the safety circuits inside the control are closed during normal operation of the furnace, but open in case of a flame-out. The contacts are wired in series in the control circuit of the unit. When they open, they shut the burner down. Once the contacts in the safety circuit open, they must be reset manually before a burner can be energized (Figure 8-9). Reset is normally by means of a button exposed on the outside of the combustion safety control panel on the unit.

Seasonal Changeover

Seasonal changeover is the change from a heating system in the winter to an air-conditioning system in the summer. Most smaller heating and cooling systems have automatic changeover or manual changeover from a switch on the thermostat (Figure 8-10), but larger systems may need to be changed over manually. Manual-changeover systems are commonly used on large hydronic systems where the recovery time of the system is relatively slow. Recovery time is the length of time required to go from heating to cooling. Seasonal changeover may be accomplished directly by the use of manual switches or through a changeover relay that is energized by a manual switch.

Direct changeover. When seasonal changeover is by use of manual switches, there will usually be a switch on either the electrical control circuitry or the line voltage to each system. The changeover is made simply by opening the switch on the system not wanted at the time, and closing the switch on the system wanted. The switches are usually similar to an electrical contactor. This

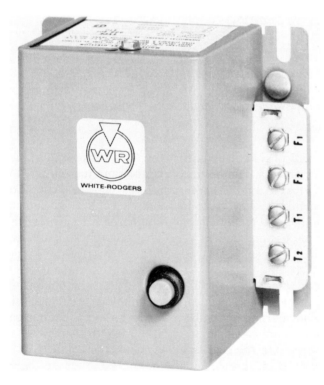

Figure 8-9 Some safety controls, such as combustion safety controls, require manual override.

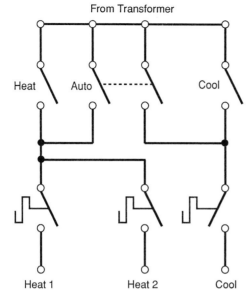

From Transformer

Heat   Auto - - - - - - - -   Cool

Heat 1   Heat 2   Cool

Figure 8-10 Seasonal changeover is often a function of the thermostat used for a given system.

system would by used only where an operating engineer is present to make the change.

Indirect changeover.   Seasonal changeover by use of a changeover control energized by a manual switch requires that a changeover relay or contactor be used (Figure 8-11). This must be a control with one set of contacts that are normally open and another that are normally closed (NONC). When the relay is energized, the contacts switch, or reverse their operation. The normally open (NO) contacts close and the normally open (NC) contacts open. One of the sets of contacts is wired in series in the control circuit to the heating system, and the other is wired in series in the control circuit in the air-conditioning system. The coil on the relay is wired in series with a manual changeover switch. When the manual changeover switch is open, the control circuit to the system wired through the normally closed (NC) contacts will be active. When the manual changeover switch is closed, the control circuit through the normally open (NO) contacts will be active.

SUMMARY

Manual controls are controls that must be set by hand. They require the presence of an operator, and they obviously are not considered as being automatic in any way. Manual controls are not often used on HVAC systems because they are not automatic, but in some cases decisions should be made by an operator or service technician before a piece of equipment is turned on or off; in these cases manual controls are used.

Summary   **139**

Figure 8-11 A system switch provides seasonal control.

Disconnect switches, used to isolate a piece of equipment while it is being worked on, are perhaps the most common manual controls used in heating and air-conditioning units. In addition, some motor control switches are also manually operated. Reset of combustion controls is nearly always manual. This is to make sure that a heating unit that fails to fire on a call for heat will not start automatically until a service technician finds out what is wrong, and corrects any faults before it relights.

In some large systems the changeover from heating control to cooling control is done manually. This is often the case in large hydronic heating and cooling systems. Because of the large amount of water in a hydronic system, some time is needed between heating and cooling cycles.

Manual Controls   Chap. 8

QUESTIONS

8-1. Name three types of manual controls used on modern heating and air-conditioning equipment.

8-2. What is a manual disconnect used for primarily?

8-3. A circuit breaker functions as both an overcurrent protection device and a disconnect switch. Which function is automatic and which is manual?

8-4. Why does a switch used for direct manual control of an electric motor need to be snap acting?

8-5. *True or false:* The magnetic coil in a contactor is always wired into the load side of the electrical circuit when used to control a motor.

8-6. What is the principal difference between a magnetic contactor and a magnetic starter?

8-7. *True or false:* The direction of rotation of all motors can be controlled by use of a voltage regulator.

8-8. How is the rotational speed of a multitap motor changed?

8-9. Why is it important to be able to change the rotational speed of a direct-drive motor used on a furnace or air-conditioner blower?

8-10. Why is it important that the reset on a combustion safety control be manual?

8-11. *True or false:* Seasonal changeover for heating and air conditioning of a building is necessary only on those systems with both heating and air-conditioning systems in the building.

8-1. Name the major parts of the electrical power distribution system shown in the accompanying diagram by writing their names in the spaces provided on the tags. Describe the part each plays in controlling the flow of electricity from its source to the point-of-use.

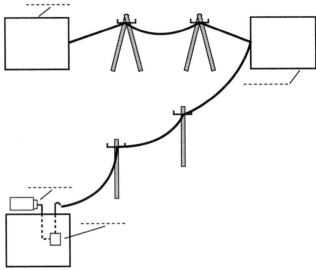

8-2. Complete the accompanying wiring diagram for controlling an electric motor, by filling in the missing wiring.

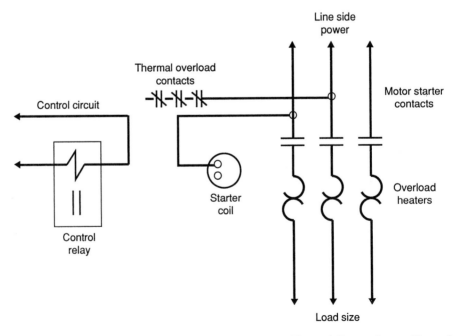

8-3. On the accompanying diagram trace out the wiring to the motor relay, marked CR, from the directional control, marked F or R, by using a red pen or pencil. Show what direction the motor would turn if wired as shown, and what direction it would turn if the directional switch was reversed.

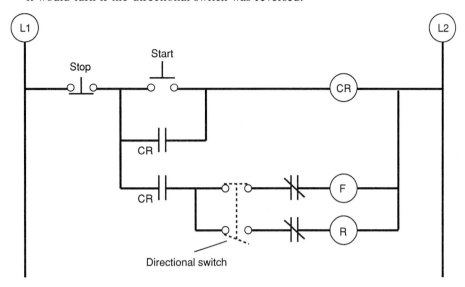

8-4. Write in the names of the major parts of the combustion safety control shown on the accompanying diagram, in the blank spaces provided. Describe the function of each of the parts in the combustion safety process.

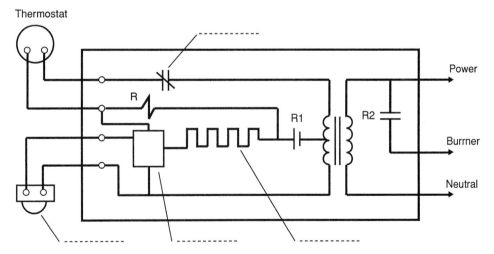

8-5. On the accompanying diagram of part of a pneumatic control system draw in the pneumatic piping required to make the system function as a seasonal changeover system.

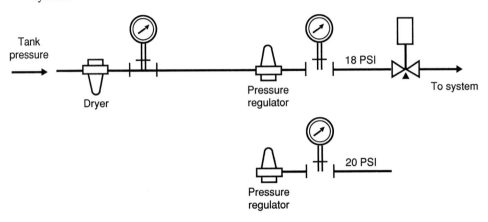

9

Electric Controls

Electric controls use electricity as a control signal and as the motive power for the control actuators. Electric controls are most commonly used as unit controls and are often used for complete control systems.

CONTROL MEDIUM

The control medium in an electric control system is alternating-current (ac) electricity. The control circuits are electrical circuits with a low-voltage power source, conductors, control components acting as switches, and operators in the actuators acting as loads. The circuit voltage may be the same as that used for power in the system (line voltage), or it may be low voltage. Line-voltage controls can be the same voltage as the system voltage, or it may be lower than system voltage but still higher than that usually considered low voltage. Low voltage used for controls is usually 24 volts.

> **Example** An air-conditioning system that runs on 480 or higher voltage may have 120-volt ac controls. In a 240-volt circuit the controls may be of the same voltage as the load voltage: that is, 240-volt or 120-volt ac controls.

Line-Voltage Controls

Line-voltage controls are usually wired so that they are directly in series with the loads in the system. In these systems the control components are sized so that they can carry the entire electrical load (amperage) required by the equipment (Figure 9-1). In other systems the line-voltage controls are used for pilot duty and do not control loads directly. *Pilot duty* means that the line voltage controls loads indirectly through intermediate controls such as contactors or starters. An example of line-voltage control is found in electric heater control. The heaters are energized by electrical power and also by electricity of the same voltage. The components of the controls have to be made heavy enough to carry the amperage required for the heaters.

Low-Voltage Controls

Low-voltage controls are used when it is desirable to have closer control than is possible with most line-voltage systems (Figure 9-2). Low-voltage controls can be made with lighter components, such as contacts, bimetallic operating elements and other parts, and thus can be made to operate with closer tolerances. The wiring for low-voltage controls can usually be of smaller gauge, and it is less expensive and easier to run between the control components. An example of a low-voltage control system is a typical gas-fired furnace control for heating. Line voltage is transformed to 24 volts dc by use of a transformer. A low-voltage circuit connects the thermostat, limit controls, and gas valve in series. The thermostat is a temperature-controlled electrical operating switch. On a call for heating, a set of contacts close in the thermostat and the gas valve is energized. When the temperature is satisfied, the contacts open and the gas valve closes. The limit controls are normally closed, temperature-actuated switches. They open only when the temperature in the furnace exceeds a set limit. The gas valve has a solenoid or magnetic pilot valve that controls the main valve on a call for heat. The pilot valve coil is wired into the control circuit from the thermostat.

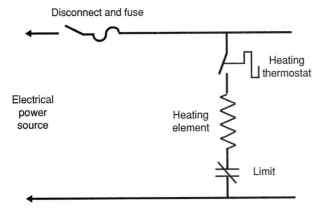

Disconnect and fuse

Electrical power source

Heating element

Heating thermostat

Limit

Figure 9-1 Line-voltage controls require that an operating control be in series in the electrical circuit with the load.

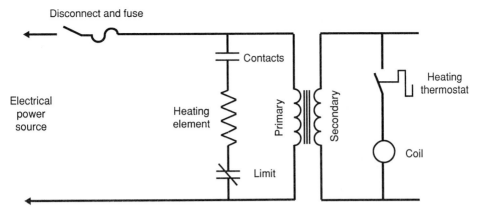

Figure 9-2 In a low-voltage control circuit a relay or other intermediate control is used to control a load circuit with a low-voltage circuit.

CONTROL OPERATION

Electric controls are controls that combine electrical and mechanical action. In most cases electricity is used as both the control signal and the operating medium; the actions of the control are mechanical. An example is the moving arm on which the contacts ride in a magnetic relay. Electricity pulls the arm in magnetically, but the arm and the contacts are mechanical parts. Electric controls are basically digital in nature. Digital means off–on. Electric controls can also be made to operate equipment in steps to give simulated modulated control, or they can be designed to operate as proportional controls to give full modulated control. *Modulated control* means to control the heating or air-conditioning equipment so that the equipment output matches the load at any given time.

Digital Control

Electric controls function primarily as off–on controls. *Digital* means "off–on" (Figure 9-3). The digital function is because electrical signals are basically digital. When an electrical controls circuit is deenergized, or the current in the control circuit is off, an "off" signal is indicated. When the circuit is energized or the current in the circuit is on, an on "signal" is indicated. This type of operation is ideal for turning devices off and on.

Stepped Controls

Stepping is the arrangement of controls to bring heating or air-conditioning equipment on in stages, or steps, to match equipment output to the load more closely. The equipment has to be built so that it can be stepped, and the controls have to be matched to the equipment (Figure 9-4). An example of a staged

Control Operation   **147**

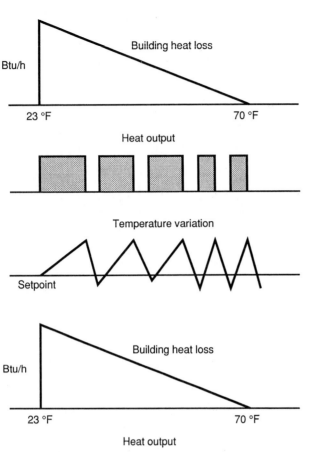

Building heat loss

Btu/h

23 °F   70 °F

Heat output

Temperature variation

Setpoint

Figure 9-3 A digital, or on–off, control circuit provides output equal to a load over a fixed period of time.

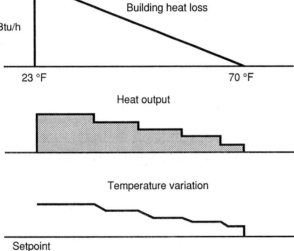

Building heat loss

Btu/h

23 °F   70 °F

Heat output

Temperature variation

Setpoint

Figure 9-4 Multistage, or stepped, control provides output in steps to match a load at a given time.

heating unit is an electric furnace with several heating elements. Each element can be controlled separately as a stage. The elements can be energized one at a time by a multistage thermostat. As each stage of the thermostat calls for heat, another element is energized.

Stepping can be accomplished in an electric control system by one of several types of control devices. Among these devices are mechanical stepped controllers and sequencers. A *mechanical stepped controller* is a mechanical device that opens or closes a series of electrical contacts in sequence (Figure 9-5). A typical stepped controller is constructed with a rotating shaft that is

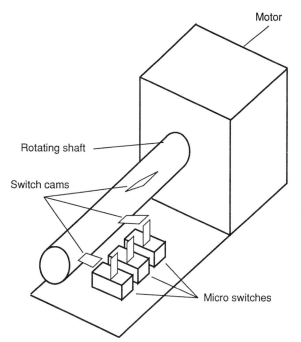

Motor

Rotating shaft

Switch cams

Micro switches

Figure 9-5 A mechanical sequencer is often used to step-control larger loads.

turned by an electric motor through a gear mechanism. This provides fairly slow rotation. The shaft has off-center cams fastened on it in such a way that they can be adjusted. There is a cam for each step in the control sequence desired. When the shaft turns, the cams turn with it, and as they turn they each activate a switch. The switches are wired in series in electrical circuits that they control. Each circuit controls one step in the control process. The motor used to turn the shaft on a stepped controller is activated by a signal from a control, such as a thermostat. When the load that the stepped controller is controlling is satisfied, the motor is deenergized and the stepped controller reverses operation. The shaft rotates back to its original position, and each switch is turned off by its respective cam operator.

A second type of stepped controller is a *sequencer.* A sequencer is similar to a multipole time-delay relay, with a delay between the opening and closing of each set of contacts. Sequencers are used to step loads on and off with a short time delay between each load. A typical sequencer includes an electric heater that is wired in series with whatever device is used to actuate it (Figure 9-6). A bimetallic element in the sequencer is heated by the heater. This causes the bimetallic element to warp, and as it warps it closes the contacts in the sequencer. When the heater is deenergized, the bimetallic element warps back to its original position and the contacts are opened in the order reverse to that in which they were energized.

Stepped controllers and sequencers are used to turn multistage heating and air-conditioning systems on and off. A large air-conditioning system may

Control Operation

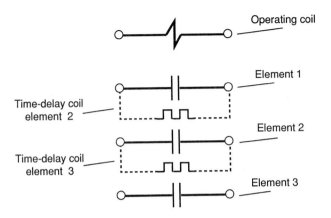

Figure 9-6 An electrical sequencer is generally used for control of smaller electrical loads.

Note: Time-delay coils to be average sensing.

have multiple compressors to provide step control. A stepped controller or sequencer could be used as the primary control to bring these on and off again upon demand from a multistage thermostat or other control.

Proportional Controls

Proportional control is control in which the output of a heating or air-conditioning unit is controlled to provide output of the equipment to match the load on the equipment at any given time (Figure 9-7). Output of the equipment in

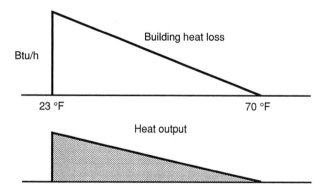

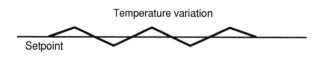

ıre 9-7 Proportional, or modulated, trol is used with equipment that can vide output that matches a load at given time.

Electric Controls   Chap. 9

a heating or air-conditioning system is the amount of heating or cooling produced momentarily. The load on the equipment is the heat load or cooling load imposed by the loss or gain of the building in which the system is installed. Most heating and air-conditioning equipment is designed to operate at full capacity when it is on. The equipment is turned off and on in short cycles to provide the amount of heating or cooling required in 1 hour's time. Modulated output is actual output of heating or cooling to match a load exactly at any time.

For a modulated control system to work, the heating and air-conditioning equipment must also be modulating. A heating unit must have a modulating burner, usually a power burner. An air-conditioning unit must also have with the ability to provide modulating output. Some type of the unloading, such as hot-gas bypass, cylinder bypass, or cylinder unloading can be used. Since an electrical control system is inherently digital, or off–on, some special effort must be made to provide modulation with electric controls. The most common way is to use a sensor that provides modulation by splitting the amperage on the two parallel legs of a bridge circuit, and then by using the split amperage to duplicate the modulating signal in the performance of an actuator. This is done through the use of special electrical circuiting called a *bridge circuit* (Figure 9-8).

A bridge circuit is actually two electrical circuits, in parallel with each other, and with two resistors in each circuit. If the resistors all have the same resistance rating, the amperage in each circuit is the same. If the resistance varies in any resistor, the amperage in the two circuits varies also. If the resistance in one leg goes down, the amperage in that leg goes up and the amperage on the other leg goes down. By using a variable resistor in place of any resistor in the circuits, the amperage in the two circuits can be controlled. If a sensor in a control system is used in place of the variable resistor, the current in the two circuits can be controlled by the system control variable.

When in use in a modulating control system, the sensor part of the bridge circuit is used as a controller (Figure 9-9). This may be in a thermostat, humidistat, or other controller used in the system. The second part of the bridge circuit, the relay that controls the motor and the feedback potentiometer, is

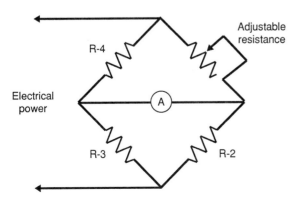

Figure 9-8 An electrical bridge circuit is the basis of the control system used with an electrical proportional control.

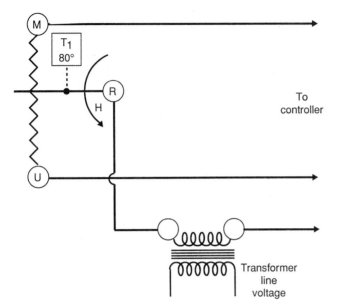

Figure 9-9 A bridge circuit is modified with half on the sensor side and half on the actuator side when it is used in a proportional control system.

part of the system actuator. This part of the system is usually incorporated with a damper or valve motor.

A form of control called *floating control* is often used to achieve the same effect as that of modulating control. In a floating-control system the rate of motion of the final control element is determined by the deviation of the controlled variable from the set point. An example of a floating-control system is the use of a digital thermostat controlling a heat motor as an actuator. A heat motor is a control comprised of a fairly large bimetallic element wound in a spiral form. The bimetallic element is fastened at the outer end of the spiral to the motor frame. The inside end of the spiral is connected to the motor shaft. An electrical heater element is wound with the bimetallic element but is insulated from it. When the heating element is energized the bimetallic spiral rotates in one direction or the other. If the heating element is deenergized, the coil rotates in the reverse direction. With the motor shaft connected to the center of the bimetallic coil, the shaft rotates with the coil and is used to operate dampers or valves.

CONTROL DEVICES

Many control devices are used in a control system, but the two primary ones are the controller and the actuator (Figure 9-10). The *controller* is a device that generates a signal in response to variation of the controlled variable from the set point. The *actuator* is a device that translates the signal from the controller into some action by the final control element.

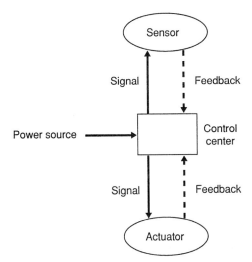

Figure 9-10 A complete control system includes a sensor, a signal, an actuator, and feedback.

Controllers

Controllers used in electric control systems are generally some type of electrical switch controlled by whatever the controlled variable is for the system. Controllers for digital and floating systems are necessarily different from those used for modulating systems. A heating thermostat acts as the controller for a typical heating system, which is a digital system (Figure 9-11). As the thermostat responds to changes in the variable controlled, an electrical switch is opened or closed. The switch is wired in the control circuit such that it will energize or deenergize that circuit.

A special thermostat must be used for a modulating system. This thermostat contains the part of a bridge circuit that generates a modulating signal.

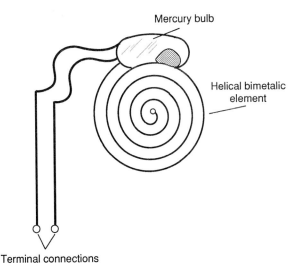

Figure 9-11 A thermostat is basically a temperature-actuated switch.

Actuators

Actuators used in electric control systems come in many different forms. Among those used for direct control are motors used to operate dampers and valves. Many others provide control indirectly. Some of these are relays, contactors, starters, solenoids, and other controls that relay on electromagnetic attraction to provide mechanical motion.

Motors.   Motors used for damper and valve operators are usually low-voltage ac motors (Figure 9-12). Generally, they are very low powered, shaded-pole motors. The shaft of the motor is connected directly to a gear train to slow the rotation rate and provide more torque on the drive shaft. The motor frame is usually built to facilitate mounting on either a damper or a valve. The electrical connections to the motor are standardized so that they can be wired into a typical control system and provide the function desired in the system.

Relays.   Most other devices used with electric control systems utilize an electromagnetic coil in some way to provide switching or proportioning action. Relays, contactors, and magnetic starters are all constructed with a set of fixed contacts and a set of movable contacts. The movable contacts are on a hinged arm that moves the contacts together or apart. The contacts are kept separated by a light spring. A magnetic coil is located so that it will pull the movable arm

Figure 9-12   A damper or valve motor is a commonly used actuator.

toward the fixed contacts so that contact is made when the coil is energized. The coil is built so that it can be actuated by either low- or high-voltage power. The contacts are made so that they will accommodate a certain amperage at a given voltage of power.

Relays are designed to operate with low-voltage draw and control only fairly low amperage (Figure 9-13). A 24-volt relay coil will normally draw about 0.12 amperes, and the contacts are designed to carry 15 to 20 amperes of current at 120 to 240 volts. Relays are generally used when it is desired to control a line-voltage circuit with a lower-voltage circuit. The control circuit is wired through the relay coil and the line voltage is wired through the contacts. When the control circuit is energized, the line circuit is closed. An example of this use is a cooling relay used to operate the fan motor on a heating/cooling unit.

Contactors.   Contactors are designed for use on applications requiring a fairly high ampere draw, from 20 amperes and up. The coil can be any voltage from 24 volts on up to 120 or 240 volts or even higher. Contactors are generally used as motor starters when the motor horsepower reaches about 1 hp or more, and when the motors have internal overload protection (Figure 9-14).

Magnetic starters.   Magnetic starters are similar to contactors, but they have built-in overload protection (Figure 9-15). The overload protection is in the form of heaters on at least one leg of the load circuit. The overload protection opens a set of contacts on either the control circuit or the load circuit in case of a current overload. The coil in a magnetic starter can be 24 volts and up, and the contactors can be for any voltage desired. Like contactors, magnetic starters come in different models for either two or three conductors. Magnetic starters are generally used for motor starters on motor applications where the motor horsepower exceeds 1 hp and the motors do not have internal overload protection.

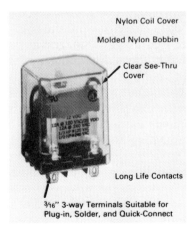

Nylon Coil Cover

Molded Nylon Bobbin

Clear See-Thru Cover

Long Life Contacts

³⁄₁₆″ 3-way Terminals Suitable for Plug-in, Solder, and Quick-Connect

Figure 9-13 A relay is an electrical control that controls one electrical circuit with another circuit.

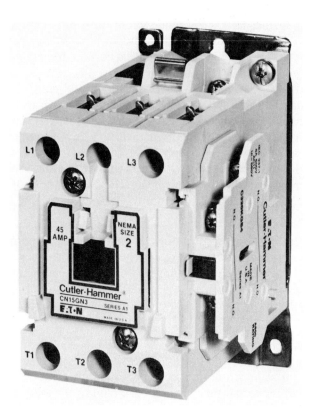

Figure 9-14 An electrical contactor is generally used to control the operation of electric motors that have internal overloads.

Solenoids. A solenoid used as an actuator operator is a magnetic coil with a movable core (Figure 9-16). The core is positioned in response to current flow in the coil. The core is used to provide the mechanical motion for the control actuator. The core in a solenoid is loose enough to move up and down freely when the coil circuit is energized or deenergized. The core is made of a ferrous material that will be affected immediately by the magnetic field when the coil is energized and yet will not hold the magnetism when the coil is deenergized. When the coil is energized and the core moves into the center of the coil, it remains there because of a magnetic field that is generated by electrical circuits generated in the coil by induction. These circuits generate a magnetic field in which the polarity opposes that in the coil, and the result is a centered core.

Solenoids are used primarily as actuators for valves to control liquid flow (Figure 9-17). Some examples are liquid line valves on refrigerant lines, and water and steam lines on heating and cooling coils. In some applications solenoids are used to operate valves directly, and in other cases they are used as pilot valves to operate lines indirectly. In the second case the pressure exerted by the liquid in the lines is used as the valve operator and the solenoid valve is used to control that pressure. An example of this application is a solenoid-controlled reversing valve on a typical heat pump system.

Figure 9-15 A magnetic starter is used to control the operation of electric motors that do not have internal overloads.

SUMMARY

An electric control system is one that uses ac electricity as a signal medium. In most cases the signal controls the equipment in a climate-control system, either directly or through a pilot control device. The signal may be either line-voltage or low-voltage electricity. Low-voltage controls normally provide closer control than do line-voltage controls.

Three different types of electric control systems are commonly used: digital, stepped, and proportional. Each control system has to be matched closely to the operating characteristics of the equipment on which it is used.

The principal mechanical devices used in an electric control system are known generically as the controller and the actuator. The controller is a device that originates a signal relative to variations between the medium controlled and set-point conditions. An actuator interprets these signals and operates equipment in the mechanical system to correct the variations.

Questions

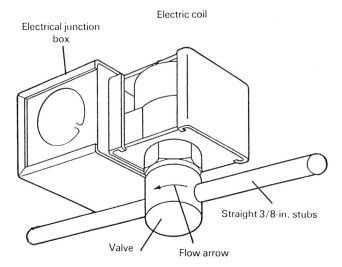

Electric coil

Electrical junction box

Straight 3/8-in. stubs

Valve

Flow arrow

Figure 9-16 A solenoid valve operator is often used to control a liquid line valve.

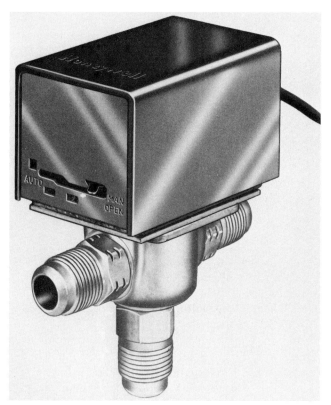

Figure 9-17 Solenoid controls are often used as valve operators.

QUESTIONS

9-1. What form of energy is used to operate the actuators in an electromechanical control system?

9-2. *True or false:* Line-voltage electrical controls that are wired in series with the loads they control need to be designed so that they can carry the full amperage of the loads in a system.

9-3. Explain what is meant by *pilot duty controls.*

9-4. Why can a low-voltage electric control system usually hold the control variable within closer tolerances than those of a line-voltage system?

9-5. Explain the term *digital.*

9-6. What relationship must exist between a control system and the type of equipment it is used to control?

9-7. Describe how a mechanical stepped controller works.

9-8. Explain what a proportional control system is designed to accomplish.

9-9. On what type of heating or air-conditioning equipment is a proportional control used?

9-10. How is floating control different from proportional control?

9-11. In generic terms, what is the control device called that generates a control signal?

9-12. In generic terms, what is the control device called that translates the control signal into some action?

9-13. Name three specific types of controllers used in electric control systems.

9-14. Name three specific types of actuators used in electric control system.

9-15. Name a control that is commonly used to control one electric circuit with another circuit.

9-16. Match the term in the first column with the phrase that best matches it in the second column by placing the letter that precedes the description in the space provided preceding the term.

    A. _____ Relay   a. built-in overloads

    B. _____ Contactor   b. large amperage draw

    C. _____ Magnetic starter   c. movable core

    D. _____ Solenoid   d. fairly low amperage loads

9-17. Which control device listed in Question 9-16 is used primarily for operating a flow control valve?

APPLICATION EXERCISES

9-1. On the accompanying wiring diagram identify the two points at which the voltage steps down, and mark them on drawing. Show by drawing connecting lines where control relay coils of one circuit controls another circuit.

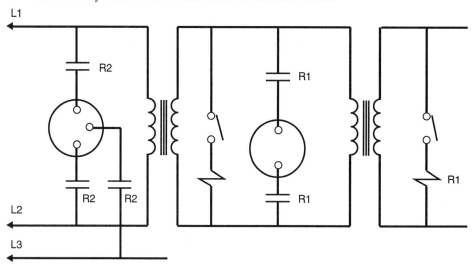

9-2. Draw in the wiring to connect the controls to the operating parts on the accompanying pictorial diagram of a furnace.

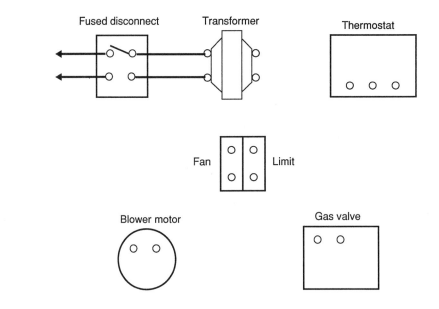

9-3. On the accompanying graph of a digital control system performance mark the points at which a furnace would be turned on and off. Also describe how the graph of the temperature in the heated space reflects this.

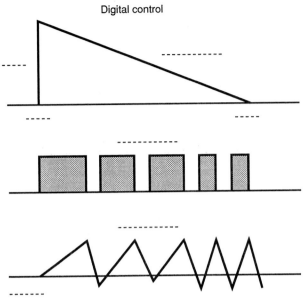

Digital control

9-4. On the accompanying graph of a stepped control system performance mark the points at which a furnace would be turned on and off. Also describe how the graph of the temperature in the heated space reflects this.

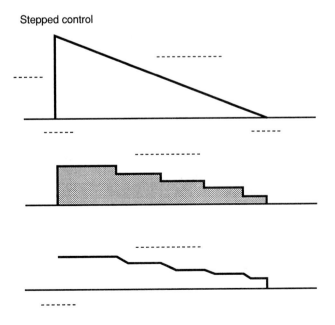

Stepped control

9-5. On the accompanying graph of the performance of a heating unit using proportional control mark the points at which a furnace would be turned on and off. Also describe how the graph of the temperature in the heated space reflects this.

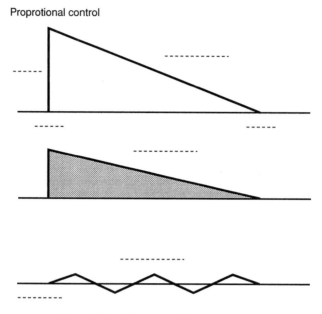

Proprotional control

9-6. On the accompanying wiring diagram of a typical furnace identify the power controls by writing the word power over them, the operating controls by drawing a circle around them, and the safety controls by writing the word safe over them.

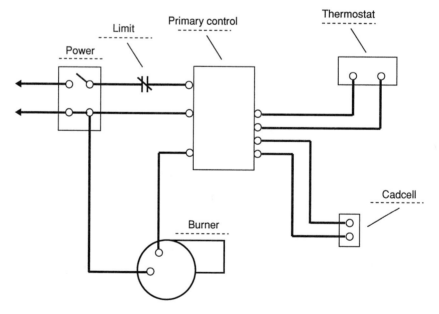

Electric Controls   Chap. 9

10

Electric Control Applications

The application of electric controls requires a knowledge of what the controls are, how they work, and perhaps most important, how to use them in a system to get the best performance out of that system. In this chapter we cover electric control application for heating, air conditioning, ventilation, and other control systems.

HEATING SYSTEMS CONTROL

Heating control systems are designed to control the heating output from a selected heating unit so that it will match the heat load for a particular building. Some heating systems operate digitally; that is, the heating element provides full output on a call for heat and no output at all when the call ends. Others are designed to provide heat in steps or stages, while others modulate output to produce heat at the same rate as it is needed. The control system must be matched to the heating equipment in such a way as to be compatible with operation of the equipment.

Digital Heating Control

Digital control systems are designed to operate any heating equipment that runs primarily in digital or on–off modes. This includes most combustion equipment with atmospheric burners, some electric heaters, and many hydronic heating units, such as fan coil units. The control signal for a digital electric control system is line- or low-voltage ac electricity. If the system has line-voltage control, the line voltage supplied to the unit is used as the control voltage. If low voltage is used, a control transformer is used to reduce the line voltage to the low voltage desired (Figure 10-1).

A controller for a digital control heating system can be any one of several types of devices. Most of them are some form of electrical switch that is actuated by temperature, humidity, pressure, flow, or any other fluctuation in a controlled variable. The main parts of a digital controller are the sensors and the switch mechanisms used. For sensing temperatures the most commonly used sensing device is a bimetallic element, an assembly of two strips of metal with different coefficients of expansion, fastened securely together (Figure 10-2). If the ambient temperate around the metals changes, the strip warps one way or the other, depending on its construction. It will warp one way on temperature rise

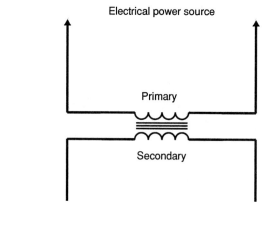

Figure 10-1 A step-down transformer used for control systems is connected to an electrical power supply on the primary side, and the voltage is stepped down on the secondary side.

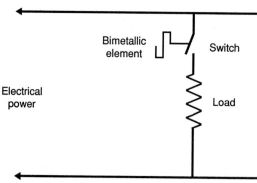

Figure 10-2 A bimetallic element is often used as a temperature sensor for control devices.

and the other way on temperature fall. The movement of the bimetallic element is used to operate the switch mechanism of the controller.

A remote-bulb thermostat is another device used for sensing temperatures (Figure 10-3). It is often used in situations where it is desirable to have the switching mechanism placed in a differnt location from that of the sensor. In the switching mechanism a remote-bulb thermostat has either a diaphragm or a bellows type of element connected mechanically to the switch. The diaphragm or bellows is connected by a capillary tube to a small bulb that is partially filled with a volatile fluid. The bulb is located where the temperature readings are desired. Any temperature variations result in pressure variations in the liquid in the bulb. These variations are transmitted to the bellows or diaphragm through the capillary tube. The switch opens or closes according to pressure variations.

Several different types of switching actions are in common use in digital heating controllers. Among the most common are switches that have sets of contacts that are opened and closed electromagnetically or mechanically by sensor action (Figure 10-4). One of the two contacts in a set is fixed, with the other movable. The movable contact is on an arm that moves back and forth to make or break contact. The arm with the movable contact is actuated by a sensing element in the controller.

A second type of switch used in a digital control system is a mercury bulb switch (Figure 10-5). A mercury bulb switch is comprised of a sealed glass bulb with a drop of mercury in it. Electrical contacts are sealed in the glass in one or both ends, depending on the type of control desired. The electrical contact leads are flexible and are connected to terminal connectors. The glass tube is

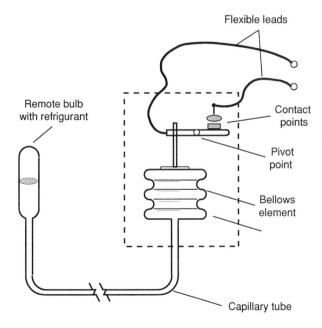

Flexible leads

Remote bulb with refrigurant

Contact points

Pivot point

Bellows element

Capillary tube

Figure 10-3 A remote-bulb control usually uses a bellows or diaphragm as the operating element.

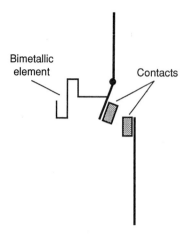

Figure 10-4 Contacts are used to make or break contact in electrical switches.

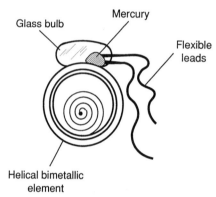

Figure 10-5 In a mercury bulb switch a drop of mercury makes or breaks contact between two electrical leads.

usually mounted on a bimetallic operating element. As the bimetallic element bends one way or the other, the tube is tilted, and the mercury ball runs from one end of the tube to the other. Electrical contact is made between the contacts in the tube when the mercury ball runs to the end in which they are located.

One of the requirements for the contacts in a switch is that they make and break quickly and cleanly. This is to make sure that the electrical current flowing in the control circuit will be turned off and on fast enough so that the contacts will not be burned by an electrical arc. To make sure of this quick break, some form of *snap-action,* or *detente device,* must be incorporated into the switch (Figure 10-6).

One type of snap action is *over-center spring action.* This is the type of action that is used in most residential light switches. In an over-center spring-acting switch the movable arm, together with its switch contact, is actuated by a spring-loaded arm. Movement by this arm is opposed by a second spring. The second spring is actuated by the switching mechanism for the control. If the switching mechanism moves the switch spring past the halfway point in its travel, the arm is snapped into place by the over-center spring.

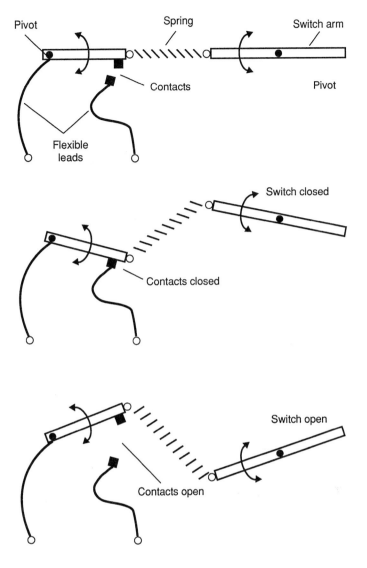

Figure 10-6 Detente is the rapid opening or closing of switch contacts to reduce arcing or welding.

To achieve snap action with a leaf-type bimetallic element switch, a small magnet is used (Figure 10-7). The magnet is located close to the bimetallic element so that it will attract the bimetallic element when it bends close to a point where the contacts make. The magnetic attraction of the magnet causes the contacts to snap closed. On opening, the bimetallic element must bend so that the contacts are drawn apart, but the magnet holds them together until they are snapped open suddenly.

To achieve snap action with a mercury bulb switch, the glass bulb is curved

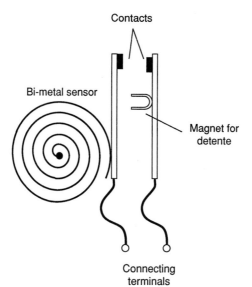

Contacts

Bi-metal sensor

Magnet for
detente

Connecting
terminals

Figure 10-7 Switches with leaf blade contacts often use a small magnet to achieve detente.

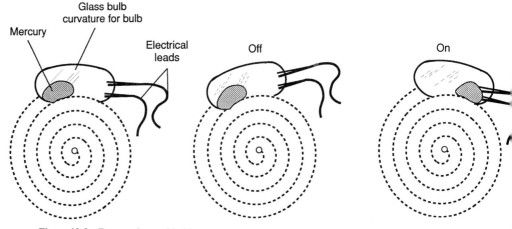

Glass bulb
curvature for bulb

Mercury

Electrical
leads

Off

On

Figure 10-8 Detente is provided in a mercury bulb switch by the curvature in the glass bulb.

slightly (Figure 10-8). The glass bulb is mounted with the curve up, on top of a helical coiled bimetallic element. To get the mercury ball to run from one end of the glass bulb to the other, the bulb must tilt far enough beyond center for the mercury to run over the center of the curve in the bulb. When it does run over the center curve it does so suddenly, and this produces the snap action.

Actuators for electric digital heating control systems are devices actuated by the control signal to perform some function to correct variations between the set point and the variable controlled. Actuators include relays, solenoid coils, motor starters, valve and damper motor actuators, and others. Relays are generally used in situations where one electrical circuit is used to control another. An example is a low-voltage control circuit used to control a line-voltage motor. Often a relay is used to operate a fan motor on a heating/cooling unit when the fan is used for both heating and cooling. Single-pole double-throw relays are often used to control two-speed motors (Figure 10-9).

Solenoid coils are used as operators for various valves and switches (Figure 10-10). Magnetic devices, they are used in some cases to operate electrical switches, but they can also be used to actuate valves. The main control in most gas valves is a solenoid-controlled valve. In some cases the solenoid operates a pilot valve that in turn controls the main valve.

Motor starters are various switches and contactors that are used to turn motors on and off. Most motor starters are similar to relays but are designed to control higher voltage power and higher amperages. Typically, motor starters are used in applications where a higher-voltage power load is controlled by a lower-voltage circuit. Motor starters called *magnetic starters* have internal over-loads on the line circuits. If a current overload occurs in the line circuit, the overload will open contacts in the control circuit and open the main contacts. Magnetic starters are normally used with motors that do not have internal overloads.

Valve actuators may be either electromagnetic or motor operated. Electromagnetic actuators are usually of solenoid type (Figure 10-11). A solenoid coil positions a stem that opens or closes the port in a valve. The valve may be used to control any liquid. They are often used for two-position control of hot water or steam for heating coils. Valve motor operators are motors that open or close a valve by rotating the stem or by moving it laterally. They are similar to damper operators that move a shaft through a discrete arc. The motors used in the operators are usually low-power shaded-pole motors. Most have gear-

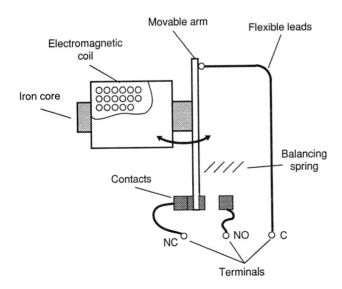

Figure 10-9 A single-pole double-throw switch is one in which one set of contacts close as another set opens.

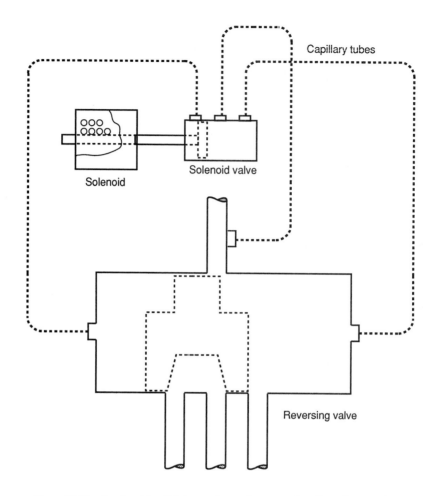

Figure 10-10 A solenoid switch is usually used to control a pilot valve on a heat pump reversing valve.

boxes as part of the motor assembly, which reduces the speed of rotation. They are bidirectional, so they can both open and close a valve (Figure 10-12).

Stepped Heating Control

Stepped control is control of mechanical equipment in steps, or stages. Each step or stage of control is basically digital, but the overall effect is similar to analog rather than digital control. Step controls are always used with heating or air-conditioning equipment that can be operated in steps, or can be unloaded to perform so that it produces a heating or cooling effect in steps or stages. Stepped controllers for electric systems come in two general types: a mechanical controller and an electrical controller.

Figure 10-11 A solenoid is often used as a valve operator.

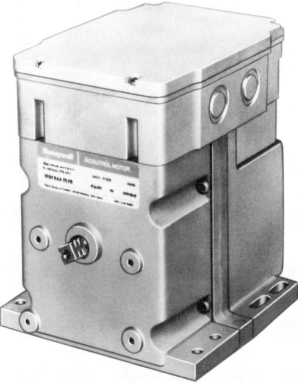

Figure 10-12 A bidirectional, or proportional, motor may be used as a damper or valve operator.

A mechanical stepped controller is a device in which electrical switches that control different steps of control are opened or closed mechanically (Figure 10-13). A typical mechanical controller has a shaft with a number of cams located along it. The cams are concentric to the center of the shaft and have electrical switches located adjacent to them. As the shaft turns the cams turn the switches on or off. The switches are turned on as the shaft turns in one direction and are turned off as it turns in the other direction. The shaft is turned by a motor through a gearbox. The gearbox allows the shaft to turn relatively slowly when the motor is turned on. Loads such as electric heating elements are controlled by the switches. Energizing the motor turns the shaft and switches the loads on in a selected sequence, and deenergizing the motor causes the switches to turn the loads off in reverse order.

An electrical stepped controller is a control device used to turn equipment off and on in steps or stages, just as does the mechanical stepped controller (Figure 10-14). Most electrical stepped controllers have a bimetallic element that turns a series of switches on and off. The bimetallic element is heated by a small electric heating element. The electric heating element is connected in the control circuit so that it is controlled by the system controller. When the bimetallic element is heated, it begins to bend. As it bends it actuates the switches in sequence. Each switch is connected, electrically, to a load, and the loads are brought on line in sequence. To reverse the action the electrical heater is deenergized. As it cools it straightens out, and the switches are opened in turn. This disconnects the loads in turn.

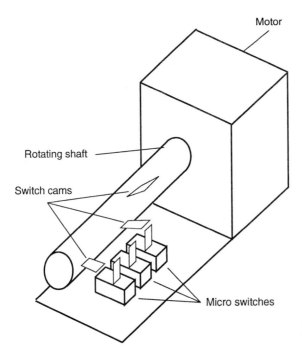

Motor

Rotating shaft

Switch cams

Micro switches

Figure 10-13 A sequencer is used to energize elements in steps.

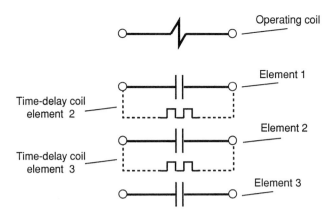

Operating coil

Element 1

Time-delay coil
element 2

Element 2

Time-delay coil
element 3

Element 3

Figure 10-14 An electrical sequencer
provides a time delay between the step
of operation.

Note: Time-delay coils to be average sensing.

Actuators used with an electric step control system are usually relays or contactors that have their contacts wired into the line-voltage circuits running to one stage of a system. A typical application is a multistage electric furnace with several electric heating elements. Each element is controlled by a relay or contactor. The stepped controller in the system operates the relays or contactors in turn, thus controlling the heat output of the unit.

Proportional Heating Control

Proportional controls are controls used to operate equipment that can be controlled so that heating or cooling output can be modulated to match the load. This requires the use of a modulating burner in the case of heating equipment, and unloading in the case of air conditioning. In the case of proportional controls the controller and the actuator must work together to provide proper control.

To achieve modulation on a combustion heating unit the unit must have a burner that allows the fire to be turned down or operated at a percentage of its full capacity. This is usually achieved by the use of a power burner. For derated output the system actuator must be able to turn down simultaneously the fuel flow and the combustion air introduced to the burner.

AIR-CONDITIONING SYSTEMS CONTROL

Air-conditioning control systems are designed to operate air-conditioning equipment to provide enough cooling to match the cooling load on a building at any time. To accomplish this, the cooling equipment output must be great enough to cool the building at maximum load conditions. The equipment is then cycled on and off to provide less than maximum output when the load is less than maximum.

Air-Conditioning System Controls **173**

Three basic types of control systems are used to operate air-conditioning equipment so that the output will match the load at any given set of conditions. The first method is digitally, or in on–off cycles. By cycling the equipment on and off in relatively short time periods the average amount of cooling effect needed over a given time can be produced. The second method is by stepped control. In stepped control the air-conditioning equipment is operated in steps. The control system brings the equipment on in steps as needed to match a load at any given time. The third method of control is proportionally. In a proportional control system the cooling output of the equipment is controlled to provide the exact amount of cooling required at any given time. In both the step and proportionally controlled systems the equipment must have a multiple-speed compressor, multiple compressors, or unloading facilities to allow its output to be controlled in steps or proportionally.

Digital Air-Conditioning Control

A digitally controlled electric control system for an air-conditioning system includes a low-voltage control power source, a cooling thermostat, and a relay in the air-conditioning equipment that controls line-voltage electrical circuits for operating the equipment. The thermostat calls for cooling by closing the control circuit if the air temperature at the thermostat rises more than 2°F above the set-point temperature. The air-conditioning equipment runs until the room temperature drops 2°F below the set-point temperature. The equipment runs in cycles to provide the amount of cooling required over a period of time to just offset the cooling load.

The sensitivity of the thermostat is increased by using a cooling anticipator. This is a small electric heater in the thermostat that is energized when the thermostat is not calling for cooling. The heater provides enough heat inside the thermostat case to cause the thermostat to call for heat slightly sooner than it would if it were affected only by room temperature. The actuator for a digital air-conditioning control system is usually a relay. The relay coil is energized by the control signal from the thermostat, and in turn, controls line-voltage circuits that are wired to the compressor and condenser motors.

Stepped Air-Conditioning Control

Stepped control of an air-conditioning system is achieved by bringing on the various steps of the air-conditioning system in response to variations of space temperature from the set point. A stepped air-conditioning control system includes a control power source, a multistage thermostat or a stepped controller, and relays or operating controls from each step of equipment operation. A multistage cooling thermostat is one that contains more than one sensor and switch.

Example A two-stage mercury bulb thermostat used to control an air-conditioning unit with two compressors. Each stage of the thermostat is controlled by a bimetallic sensor and a mercury bulb switch. The two sensors are arranged so that there is a 2°F difference between the temperatures required to make and break the circuits related to them. The first stage of cooling is called for if the space temperature rises 2°F above the space temperature. The second stage "makes" if the space temperature rises another 2°F, or 4°F above the set-point temperature. The two stages of the thermostat are wired in separate control circuits, each controlling a relay or other control device, operating one step of the cooling system.

Proportional Air-Conditioning Control

Proportional control of an air-conditioning system provides cooling output that is proportional to the cooling load on a building at any given time. An electric proportional control system consists of a control system power source, a thermostat that can sense variations between controlled variable conditions and set-point conditions, and an actuator that can operate an air-conditioning system in such a way as to provide output as required for varying loads. To accomplish this, the air-conditioning system must include unloaders. The methods of unloading a system are hot-gas bypass, cylinder bypass, and cylinder unloading.

Hot-gas bypass is the bypassing of some of the hot refrigerant from an air-conditioning compressor around the system so that all of the refrigerant does not go through the system condenser or the evaporator (Figure 10-15). This is done by placing a piping bypass in the hot-gas refrigerant line as it leaves the compressor, and running the bypass directly to the suction line. A three-way valve is used where the bypass line leaves the hot-gas line, to control

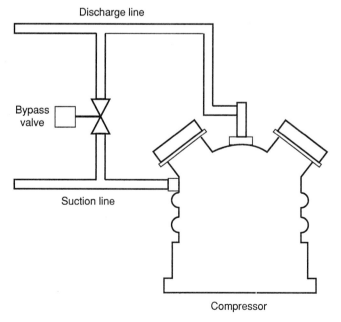

Figure 10-15 One method of unloading an air conditioner is by use of a hot-gas bypass system.

the flow of refrigerant around the compressor and to unload the evaporator. The three-way valve is operated by the control system to provide the unloading necessary to provide just the amount of cooling required for a building.

Cylinder bypass is achieved by piping the discharge gas from some of the cylinders of a multicylinder compressor so that the discharge from them is circuited back to the suction side of the compressor (Figure 10-16). Cylinder bypass piping is usually installed in the compressor at the factory and is not a field installation. Bypass of the refrigerant is controlled by a three-way valve in the bypass line. The valve is operated by the controller in the control system to provide the amount of unloading needed.

Cylinder unloading is mechanical control of the discharge valves on some of the cylinders of a multicylinder compressor (Figure 10-17). When energized by the control system, the unloading mechanism holds the valves open so that the cylinders they are on will not function as operating cylinders. A mechanical or hydraulic valve lifter is used that holds the valves open. The effect is to allow only some percentage of the refrigerant to move through the system. The cylinder unloaders are operated by the controller in the control system to provide the amount of cooling needed at a given time.

Controllers used in proportional control systems must be matched to the equipment used. Electric controls are inherently digital in operation. To achieve

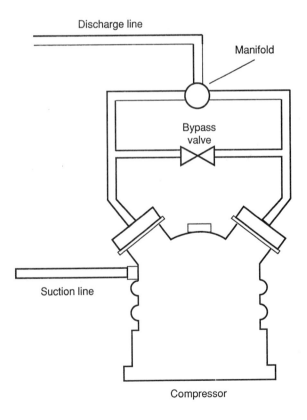

Figure 10-16 Cylinder bypass is used in some cases for unloading an air conditioner.

Electric Control Applications   Chap. 10

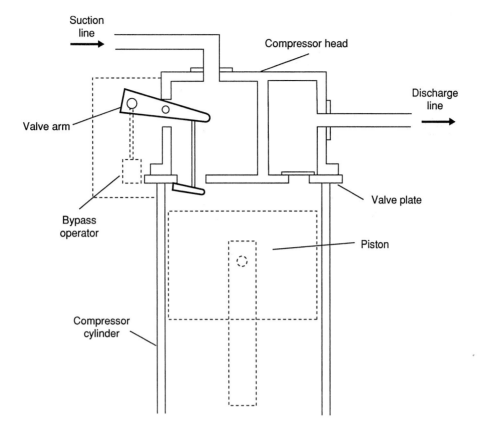

Figure 10-17 Cylinder unloading causes a cylinder in a reciprocating compressor to become inactive.

proportional control with electric controls requires the use of an electrical controller that provides a variable electrical signal in response to variations in the variable controlled. The controller can be a thermostat, pressurestat, or other control related to the system being controlled.

Most electrical proportional control systems employ some form of bridge circuit in which parallel circuits are used to indicate variances from set point (Figure 10-18). The sensor in a controller generates a split circuit signal. The signal is interpreted by an actuator, to provide control proportional to variables.

HUMIDITY CONTROL

Most humidifiers are mechanical devices that produce water vapor in some form that is injected into the supply airstream of a building. To increase the humidity in the building the equipment can be operated in an on–off manner. A digital control system is generally used. Either a line- or low-voltage humidistat senses

Humidity Control

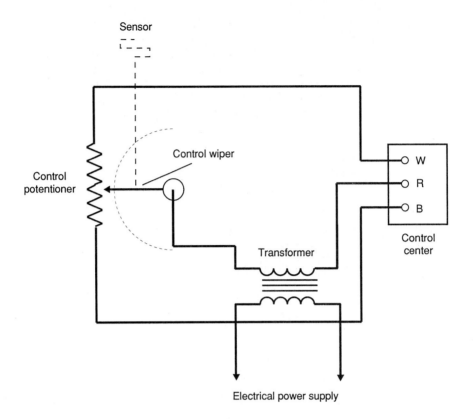

Sensor

Control wiper

Control
potentioner

Transformer

W

R

B

Control
center

Electrical power supply

Figure 10-18 A typical proportional control system using a bridge circuit for control is a three-wire system.

the humidity level in the building, and a switch in the humidistat is closed or opened by variations between the actual humidity and the set-point humidity. A relay or contactor in the humidifier or simply as part of the control system opens a water valve and turns on any other mechanical devices necessary to provide the humidity.

> **Example** In one type of humidifier a water valve controls water flow through a relatively small nozzle (Figure 10-19). The stream of water from the nozzle is directly against a disk that is rotated rapidly. The rotating disk breaks the water stream into very small droplets and directs them into the space to be humidified. The water valve and a relay to turn the motor on and off are controlled by a humidistat in the space.

DAMPER CONTROL

Both digital and proportional control systems are used for damper control. Usually, the dampers are used on outside air for ventilation of a building. If

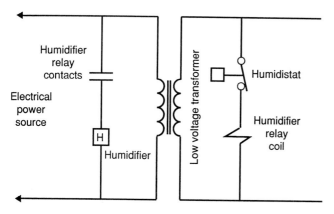

Figure 10-19 In a humidifier, water flow is controlled by a humidistat.

the damper system is very simple, a digital control system would generally be used with the damper normally open and a remote-bulb thermostat located downstream from the damper to act as a low-temperature limit (Figure 10-20). If a more sophisticated damper system is used, the outdoor air may go into a mixed-air plenum with a return-air duct also entering it. A proportional control system would be used with a remote-bulb sensor in the mixed-air plenum, and the dampers would float between open and closed to provide a desired mixed-air temperature. The mixed air would then enter the heating and/or cooling section of the system.

SUMMARY

Electric control systems use ac electricity as a control signal. Controllers are basically electrical switches operated by variations between actual conditions and set-point conditions. Electrical circuits carry the control signal, and electromechanical devices are used for actuators. There are three basic types of electric control systems: digital, stepped, and proportional.

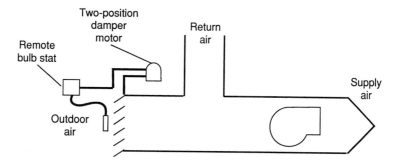

Figure 10-20 An outdoor air controller operates a damper tool to control outdoor air flow.

Digital control is straight on–off control. The controller in the system either calls for action or it does not. The actuator is either on or off. Stepped control is control in which the operating equipment is brought on, and turned off, in steps or stages. Use of this control system presupposes that the equipment can be operated in steps. An electric heating unit that has more than one heating element is an example. Proportional control is control in which system output is controlled in such a way that output always matches load. Like a stepped control system, proportional control is used with equipment in which output can be modulated.

Humidity control is usually by way of a digital control system. Control of either a humidifier or a dehumidifier can best be achieved by on–off control. Damper control can be either by digital or proportional control. If a damper system is very simple, on–off control is probably called for. If the system is more complex, perhaps including an economizer system, a modulating control system is more common.

QUESTIONS

10-1. *True or false:* It is necessary that a control system be matched with the mechanical equipment it is controlling.

10-2. *True or false:* A digital control system is used primarily for controlling equipment that operates in on–off modes.

10-3. Describe a bimetallic control sensor.

10-4. What is the most outstanding characteristic of a remote-bulb thermostat?

10-5. Name two types of switches commonly used in electric control system controllers.

10-6. Name three different ways in which detente, or snap action, is achieved in control switches.

10-7. Name five types of actuators used in electric control systems.

10-8. *True or false:* Stepped control could be called a series of digital control actions.

10-9. Name two types of stepped controllers used in electric control applications.

10-10. Name three ways in which capacity modulation can be achieved on an air-conditioning system.

10-11. *True or false:* A proportional control system has an actuator that produces a signal that is proportional to variations between the controlled variable and the set point.

APPLICATION EXERCISES

10-1. On the accompanying diagrams of different types of switches used in control devices identify the feature that makes each switch snap acting, and explain how the feature provides snap action.

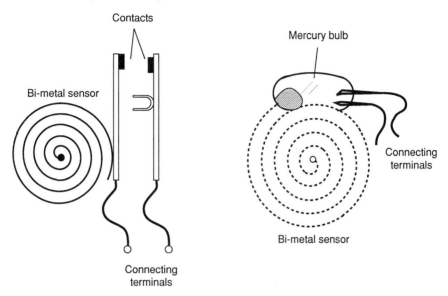

10-2. Complete the following wiring diagram of an air conditioning control system with a liquid line solenoid valve used in a pump down control system. Explain how a solenoid operator compares to an electric relay operator.

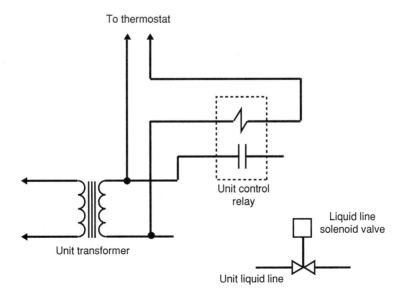

10-3. Complete the accompanying wiring diagram to show the control voltage wiring of a step controller on an electric furnace. Describe the operation of the step controller in bringing on the furnace elements one at a time. How will they then step off?

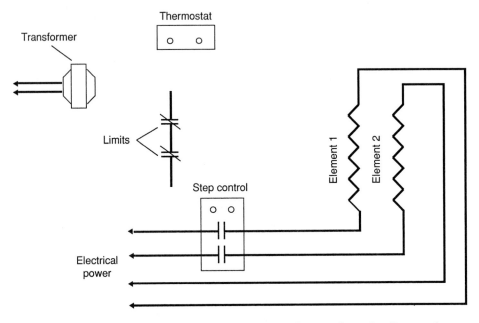

10-4. Complete the accompanying refrigerant piping diagram for unloading an air conditioning system compressor by drawing in the missing sections of the diagram.

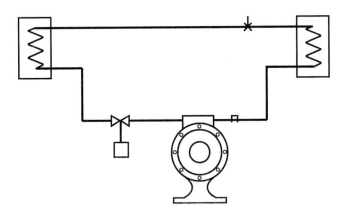

10-5. On the two diagrams of a typical electrical humidistat draw in the position of the contacts for first, the humidistat calling for humidity, and second, with the humidity satisfied.

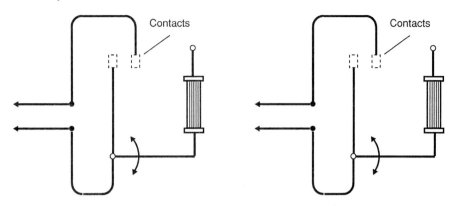

10-6. On the accompanying diagram of a damper system what would the dry bulb temperature of the mixed air be with the quantities and temperature shown for outside air and return air?

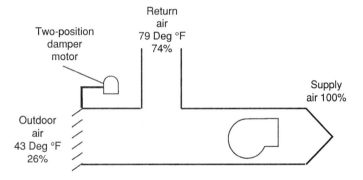

11

Pneumatic Controls

Pneumatic controls use relatively low-pressure air as a control medium. The control signal consists of variations in air pressure. In most cases air pressure is also the motive power for the control actuators. A pneumatic control system is inherently a proportional system. Pneumatic controls have been used for many years for controlling the parts and components of heating and air-conditioning systems.

SYSTEM

In a pneumatic control system, clean dry air is used as a control medium. The main parts of a pneumatic control system are the compressor and air tank for supplying air to the system, tubing or piping for carrying the air, an air drier and filter, a pressure regulator, controllers, and actuators (Figure 11-1). The controllers generally function as variable-pressure regulators. A sensor is used to sense variations in temperature, humidity, air pressure, or other control variables, and the controller sends out a variable air pressure signal to the actuator in a system. The actuators operate various parts of the heating or air-conditioning systems.

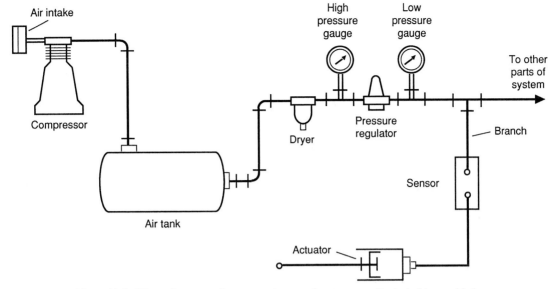

Figure 11-1 The major parts of a pneumatic control system are all selected to provide low-pressure air as a control medium.

Pneumatic Control Medium

The control medium in a pneumatic control system is low-pressure (0 to 20 psi) air. To provide the medium, ambient air is compressed and stored in a pressure tank (Figure 11-2). Air is piped from the tank to the other control components in the system through copper or plastic tubing or small pipes. To ensure that the air is clean and dry, it goes through a filter/drier as it leaves the tank. The

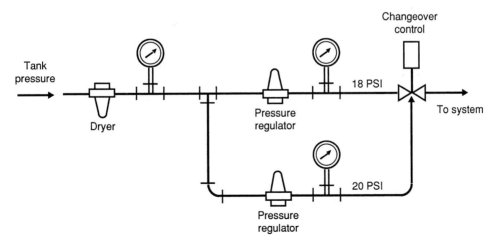

Figure 11-2 Pressure regulators in the main supply line of a pneumatic control system regulate the supply air pressure.

air goes through a pressure regulator, where the pressure is reduced to that desired in the main supply lines, usually about 18 to 20 psi. The air then goes to the various control circuits in the system through tubing or piping. The control system may be made up of all pneumatic controls, or it may have pneumatic controls interfaced with electric or electronic control elements.

PNEUMATIC CONTROL OPERATION

Pneumatic controls are basically analog in nature. *Analog* means that the controller will provide a proportional signal in response to changes in the control variable and the operator will respond with a proportional effect. Pneumatic control systems can also be designed to operate equipment digitally or in steps. There are two basic types of controllers: direct acting (DA) and indirect acting (RA) (Figure 11-3). In a *direct-acting controller* the control air pressure goes up as the ambient temperature goes up. In an *indirect-acting controller* the control air pressure goes down as the ambient temperature goes up. Ambient temperature is the temperature of the air at the same location as the controller.

There are also two basic types of actuators (Figure 11-4). These are normally open (NO) and normally closed (NC). A normally open actuator is one in which the valve is open when there is no pressure on the valve operator. A normally closed actuator is one in which the valve is closed when there is no pressure on the operator.

Example For proper control of the flow of water through a heating coil the controller must be matched with the actuator (Figure 11-5). If it is desired to have no water flow when the controller is satisfied, a normally closed actuator is used with a direct-acting controller. If full water flow is desired when the controller is satisfied, a normally open actuator is used with a direct-acting controller.

Typically, a pneumatic control system is designed so that the actuators will

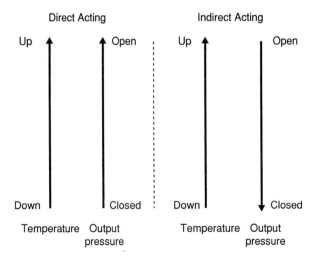

Figure 11-3 A pneumatic controller is considered to be direct acting if output control pressure goes up when the controlled variable goes up, and is indirect acting if the control pressure goes down when the controlled variable goes up.

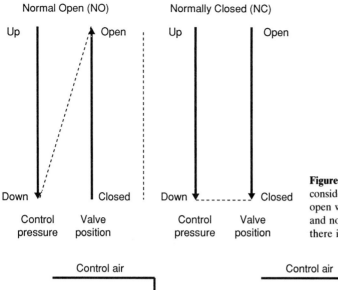

Figure 11-4 A pneumatic actuator is considered to be normally open if it is open when there is no control pressure, and normally closed if it is closed when there is no control pressure.

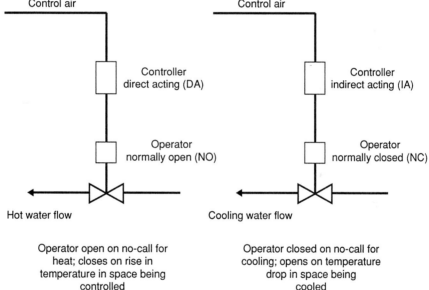

Operator open on no-call for heat; closes on rise in temperature in space being controlled

Operator closed on no-call for cooling; opens on temperature drop in space being cooled

Figure 11-5 It is necessary to match the action of the controller to the open/closed characteristics of the actuator to ensure proper system operation.

work through a range of pressures from 3 to 15 psi. That is, with a direct-acting controller the pressure will be 3 psi when the temperature is at the bottom of its control range and will be 15 psi when the temperature is at the top of its control range. The control range is matched with the temperature range. The sensitivity of the controller is the ratio of the temperature range to the control pressure range (Figure 11-6). Typically, a temperature difference of 1°F will effect a control pressure change of 2½ psi.

The actuator in a pneumatic control system is also adjusted so that it will

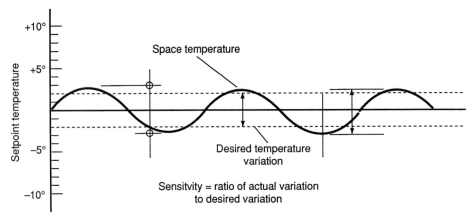

Figure 11-6 Sensitivity is the relationship of the control pressure output to variation between the set point and controlled variable.

go through a full travel span in the pressure range of the system. If a controller has an output range of 3 to 15 psi, a normally closed actuator should travel from full closed to full open in that same range. To achieve optimal efficiency with a pneumatic control system, the valve or other operator should be adjusted to be at midrange of its operation when the pressure is at the midrange (Figure 11-7). For a system that operates in the range 3 to 15 psi, the operator should be at midposition when the pressure is 9 psi.

Digital Control

Pneumatic controls function primarily as analog controls. Output from the controller closely follows changes in the control variable. Control air pressure is varied or modulated as the control variable varies. This type of operation is ideal for controlling heating or air-conditioning systems so that their output closely matches the heating or cooling load on a building, but there are occasions when it is desirable to have a pneumatic system produce a digital signal. An

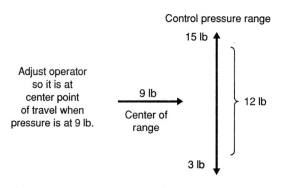

Figure 11-7 An actuator should be adjusted so that it is at midpoint in its travel range when control pressure is at midpoint of its range.

Pneumatic Controls   Chap. 11

example is when any piece of HVAC equipment that operates only in response to a digital signal needs to be operated by the pneumatic system.

To convert varying air pressure to a digital signal requires the use of a device called a *transponder.* A transponder is any device that converts one form of signal to another. A transponder that converts a pneumatic signal to a digital electrical signal is called a *pneumatic–electric switch* (Figure 11-8). In a pneumatic-electric switch, air pressure from the pneumatic system opens and closes a set of electrical contacts to control an electrical circuit. The electrical signal is then used to operate some part of the system in an electromechanical mode.

Stepped Control

Stepped control is a control arrangement that brings heating or air-conditioning equipment on in steps or stages (Figure 11-9). This is done to more closely match equipment output to the load. The equipment must be manufactured so that it can be controlled in steps, and the controls then have to be matched to the equipment.

Examples. Stepped control of an electric furnace that has several heating elements is accomplished by energizing each element in turn by use of a stepped controller. The elements are energized one at a time by a stepped controller or sequencer. As each stage of the controller or sequencer calls for heat, another element is energized.

Stepped control of multiple compressors in an air-conditioning system is accomplished by bringing the various compressors on in turn as a stepped controller calls for them. Stepped control is accomplished in a pneumatic control system by use of a control that uses a pneumatic signal for operation, but opens

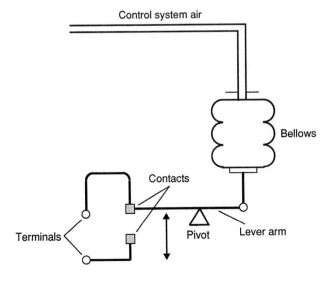

Figure 11-8 By use of a pneumatic–electric switch a pneumatic control system can be interfaced with an electric system.

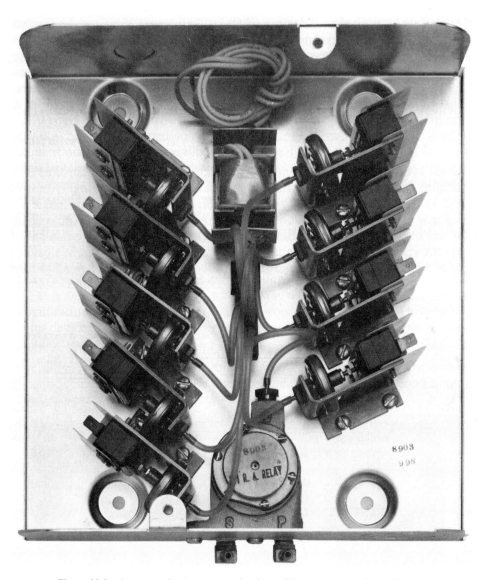

Figure 11-9 A pneumatic sequencer makes it possible to control a piece of equipment in stages or steps.

and closes electrical switches for the separate steps. Pneumatic stepped controllers are available that control up to 40 steps.

A stepped controller is a mechanical device that opens or closes a series of electrical contacts in sequence. A typical pneumatic stepped controller is constructed with a mechanical shaft that either moves back and forth or rotates in response to changes in control air pressure. The shaft has off-center cams fastened on it in such a way that they can be adjusted. There is a cam for each step in the control sequence desired. When the shaft moves or turns, the cams move with it, and as they move they each activate a switch. The switches are wired in series in electrical control circuits to the mechanical equipment that they control. Each circuit controls one step in the control process.

Proportional Control

Proportional control is control in which the output of a heating or air conditioning unit is controlled to provide output of the equipment to match the load on the equipment at any given time (Figure 11-10). Output of the equipment is the amount of heating or cooling produced momentarily. The load on the equipment is the heat load or cooling load imposed by the loss or gain of the building in which the system is installed.

Most heating and air-conditioning equipment, with the exception of hydronic equipment, is designed to operate at full capacity when on. The equipment is turned on and off by the control system in short cycles to provide the amount of heating or cooling required in any given period of time. Modulated output is actual output of heating or cooling to match a load at any time.

Proportional control by a pneumatic control system and hydronic heating or air-conditioning equipment is inherent in both the system and the equipment. For a proportional control system to work with a pneumatic control system, but with heating or air-conditioning equipment designed to operate digitally, the equipment must be capable of operating modularly. A heating unit must have a modulating burner, usually a power burner. An air-conditioning unit must also have the ability to provide modulated output. Modulation on an air-conditioner means it must be capable of being unloading. Unloading can be by hot-gas bypass, cylinder bypass, or cylinder unloading.

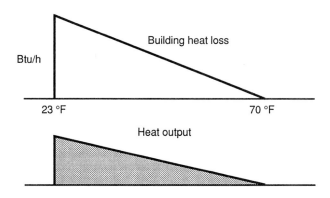

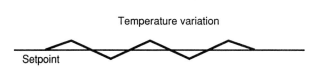

Figure 11-10 A pneumatic control system is inherently a proportional control system. As such, it provides excellent control of heating and air-conditioning systems.

A form of control called *floating control* is basically the same as proportional control. In a floating-control system the rate of motion of the final control element is determined by the deviation of the controlled variable from the set point. An example of a floating-control system is the use of a pneumatic controller operating a proportional damper motor or valve operator (Figure 11-11).

CONTROL DEVICES

The two main control devices used in a pneumatic control system are controllers and actuators. Controllers are devices that generate a signal in response to variations of the controlled variable from the set point. Thermostats, humidity controllers, and pressure controllers are some typical pneumatic controllers. A pneumatic actuator is a device that translates the signal from a controller into some action by the final control element. Damper actuators and valve actuators are typical operating devices.

Figure 11-11 A piston operator is a typical pneumatic system actuator.

Pneumatic Controls   Chap. 11

Controllers

Controllers used in pneumatic control systems are generally a form of variable-pressure regulator operated by temperature, pressure, or other control variable (Figure 11-12). A heating thermostat is the controller for a typical heating system. As the thermostat responds to changes in the controlled variable, output air pressure from the thermostat varies. The output air is piped to an actuator. The actuator is the controlled device for the control system. Special thermostats are used for different pneumatic control system operations. One type of thermostat is used for heating, and a slightly different one for cooling. Two-stage operation requires a different thermostat, as does night-set-back operation (Figure 11-13).

Actuators

Actuators, or operators, used in pneumatic control systems come in three basic forms: piston, diaphragm, and bellows. Each type has features that make it more useful for certain applications than for others. Piston actuators are generally used for operating dampers, diaphragm operators for operating valves, and bellows actuators for valves and various types of switch operators.

Piston operators. A *piston operator* is a cylinder with a piston inside that is positioned by air pressure (Figure 11-14). One end of the cylinder is sealed and has an air hose connector attached. The other end has a connecting

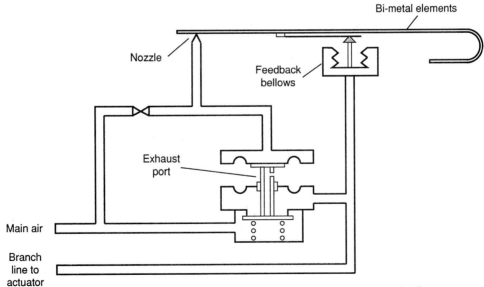

Figure 11-12 Air pressure leaving a pneumatic controller is controlled by a sensor that opens or closes a bleed port.

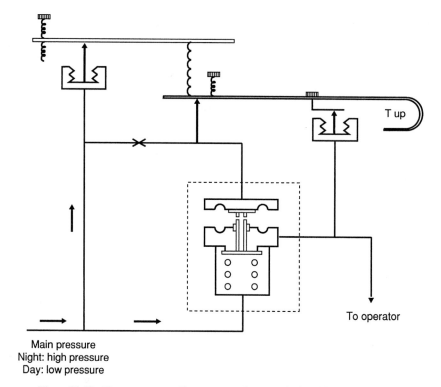

Main pressure
Night: high pressure
Day: low pressure

Figure 11-13 Two-stage control by a pneumatic controller is achieved by regulating the supply pressure to the controller.

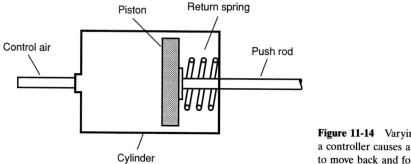

Figure 11-14 Varying air pressure from a controller causes a piston operator rod to move back and forth.

rod from the piston extending from it. There is a pliable seal between the piston and the inside periphery of the cylinder. The piston is spring returned. A spring inside the piston works against air pressure. When air pressure is introduced into the cylinder through the air-hose connector the piston is pushed against the spring and moves. When the pressure is reduced, the spring returns the piston to its resting position.

Diaphragm operators. A *diaphragm operator* is a device with a flexible composition or metal diaphragm sealed across an air chamber (Figure 11-15).

Air pressure can be exerted on one side of the diaphragm through a hose connector. The other side of the diaphragm moves a rod back and forth. The rod is used to operate a controller. One of the useful properties of a diaphragm operator is that the force exerted by the diaphragm is a function of the area of the diaphragm multiplied by the air pressure. Much pressure can be brought to bear by use of an operator with a large diaphragm area (Figure 11-16).

Bellows operators. *Bellows operators* are devices using a bellows for operating pressure (Figure 11-17). A bellows is a cylinder made of flexible metal or plastic formed with ridged, collapsible sides. The cylinder is usually formed

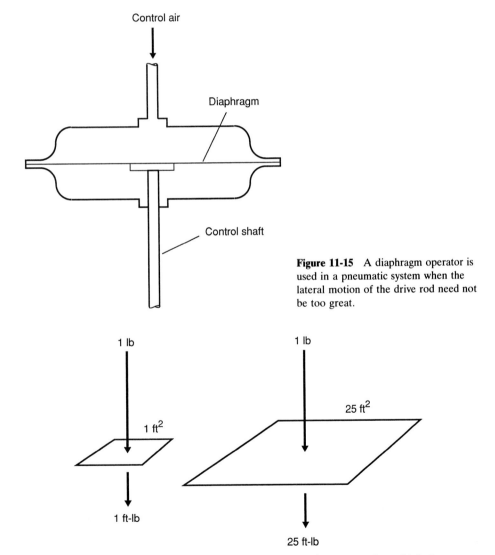

Figure 11-15 A diaphragm operator is used in a pneumatic system when the lateral motion of the drive rod need not be too great.

Figure 11-16 The force exerted by a pneumatic control system can be multiplied by making the area of the controller element larger.

Control Devices

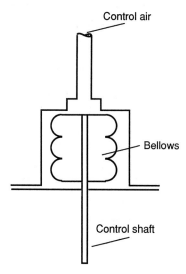

Control air

Bellows

Control shaft

Figure 11-17 A bellows operator used in a pneumatic control system is similar to a diaphragm operator except that it has more latitude of rod motion.

of thin brass or other material that will not develop cracks under continual flexing. The ridged cylinder is inside a closed can, and the cylinder is sealed around the edge. A connecting rod runs from inside the bellows through a hole in the end of the container. Air pressure is introduced into the container through a hose connector on the can. The introduction of air causes the bellows to collapse partially, moving the rod away from the container. A reduction of the air pressure in the container causes the bellows to expand, moving the rod toward the container. The rod running from the container provides operating power for a valve or switch.

SUMMARY

A pneumatic control system is one in which low-pressure air is used as a control medium. Compressed air goes through filters and driers and a pressure regulator. It then goes to the various controllers in the system, where the output pressure is modulated to indicate controlled-variable conditions. The output air is piped to actuators that operate HVAC equipment in the mechanical system.

A pneumatic control system is inherently a proportional system. A typical pneumatic controller provides an output signal that varies with variation in the controlled variable from a set point. However, a proportional system can be designed to operate equipment in steps. To do so requires intermediary control devices that convert the naturally proportional control signal to either a digital or a stepped control signal.

QUESTIONS

11-1. What control medium is used in a pneumatic control system?

11-2. Name six of the eight main components found in a pneumatic control system.

11-3. What is the usual pressure range in which a pneumatic control system operates?

11-4. Describe *modulating control.*

11-5. It is _____(necessary/unnecessary) for the heating or air-conditioning equipment used with a modulating control system to have the capability of being modulated.

11-6. *True or false:* Since a pneumatic control system is inherently an analog system, it cannot be used for on–off control.

11-7. Define the term *stepped control.*

11-8. *True or false:* Since a pneumatic control system is inherently a proportional system, it is not possible to achieve stepped control with it.

11-9. What is the generic term for the control device that generates a control signal in response to change in control variable?

11-10. What is the generic term for the control device that translates a control signal into action to correct deviation from set point?

11-11. Explain the following statement: A pneumatic thermostat can be called a variable-pressure regulator controlled by temperature.

11-12. *True or false:* One type of pneumatic thermostat serves for any type of heating or air-conditioning application.

11-13. Describe the construction of a piston-operated pneumatic operator.

11-14. *True or false:* Diaphragm operators as used in pneumatic control systems can only be used on very small valves or dampers because of limitations related to the pressure they can exert.

11-15. Two types of pneumatic operators are named in Questions 11-13 and 11-14. What is the name of a third?

APPLICATION EXERCISES

11-1. Complete the drawing of the pneumatic control system on the following diagram. Name each of the controls and parts of the system by writing in the names on the tags on the diagram. Explain how the system provides a control function by variations in control air pressure.

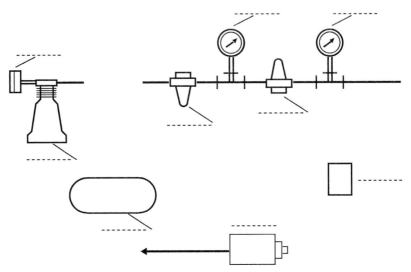

11-2. Fill in the blanks on the accompanying chart to show whether the valve used for the stated purpose should be normally open or closed.

Controller/Actuator Mix

Cooling

Controller	Valve action
Direct acting	_____
Indirect acting	_____

Heating

Controller	Valve action
Direct acting	_____
Indirect acting	_____

11-3. If a pneumatic controller is adjusted to have a sensitivity ratio of 2.25, for a temperature range of 25°F what pressure range is it set to cover?

11-4. Complete the pneumatic piping and the control wiring on the accompanying diagram to properly operate the electrical valve motor with the pneumatic system.

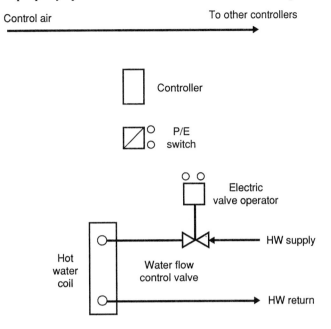

11-5. Using the air pressure and the spring pressure shown on the pressure regulator in the accompanying diagram, what would the leaving air pressure be? Trace the path of the air through the regulator with colored pen or pencil.

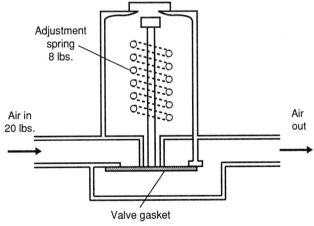

11-6. If a pneumatic piston operator has a 5 in. dia. piston, and air pressure of 6.5 lbs/ sq/in. is exerted on it, how much pressure will the piston exert?

12

Pneumatic Control Application

The proper application of pneumatic controls requires a knowledge of what the controls are, how they work, and perhaps most important, how to use them in a control system to get the best performance out of that system. In this chapter the application of pneumatic controls to heating, air-conditioning, and damper systems is covered.

HEATING SYSTEM CONTROLS

Heating control systems are designed to control the heating output from a selected heating unit so that it will match a specific heat load. The *heat load* is the heat loss of the building, including the ventilation load, at a given set of conditions. Three basic types of heating control systems are used: digital, stepped, and proportional control (Figure 12-1).

Digital Heating Control

A digital heating control system is basically one in which the heating equipment in the mechanical system is controlled by cycling it on and off to produce enough

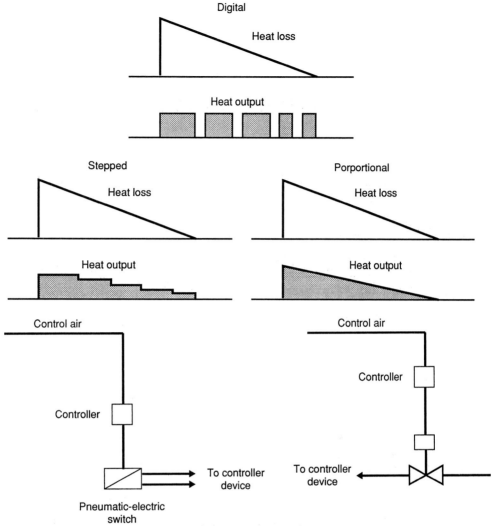

Figure 12-1 A pneumatic control system is basically a proportional system, but by using additional control components digital or stepped control can be achieved.

heat during a given time period to match the heating load during the same period (Figure 12-2). Digital control with a pneumatic control system is usually achieved by the use of a pneumatic–electric switch. A pneumatic–electric switch is an electrical switch that is actuated by control pressure from an controller in a pneumatic system (Figure 12-3). A pneumatic–electric switch is used to turn a typical heating unit off and on in response to varying pressure signals from the controller.

Example A pneumatic control system is used in a heating system using gas-fired duct heaters located on a common supply-air system. If the air temperature

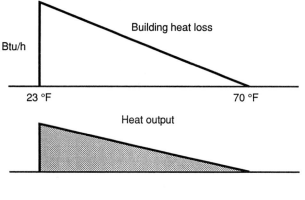

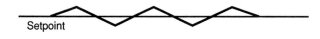

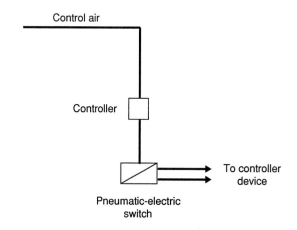

Figure 12-2 By the addition of a pneumatic–electric switch to a pneumatic control system digital control provides the average system output required to satisfy control conditions over a period of time.

in a space heated by a duct heater drops 2°F below the set-point temperature on the thermostat (pneumatic controller), the duct heater for that space should be turned on. If the air temperature in the space goes 2°F above the set-point temperature, the unit should be turned off. A pneumatic–electric switch is controlled by a pneumatic room thermostat. As the temperature in the space varies, the thermostat in the space produces a varying control output pressure. This pressure closes the electric switch in the pneumatic–electric switch when the control pressure drops enough to indicate a 2°F temperature drop. Conversely, a 2°F temperature rise in the room opens the switch. The pneumatic–electric switch closes to activate the heater on temperature drop and opens to turn it off on temperature rise.

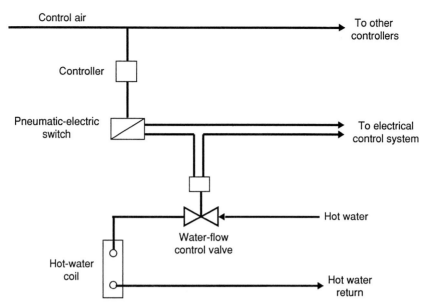

Figure 12-3 A pneumatic–electric switch provides an interface between pneumatic and electric control systems.

Stepped Heating Control

With a pneumatic system, stepped control is achieved by using a pneumatic stepped controller that is controlled by control output pressure generated by a thermostat or other primary controller (Figure 12-4). The stepped controller contains a diaphragm or bellows power head and a series of electric switches. Control air pressure from the thermostat in the system pushes against the diaphragm or bellows and closes or opens the electrical switches in sequence. The electrical switches are wired into the control circuits to step controllers in the heating unit used.

Proportional Heating Control

Proportional control of a typical hydronic heating system is achieved by using a heating thermostat, which provides varying control air pressure as the temperature in the spaces being controlled changes (Figure 12-5). The control air output from the thermostat activates a modulating valve controlling hot water flow into a heating coil. In a typical application, if the temperature in the controlled space drops below the set point for the controller, the valve starts to open. This allows more hot water to flow through the coil. If the temperature in the space rises, the valve starts to close and less hot water flows through the coil.

To achieve the proper control, a controller must be adjusted to produce

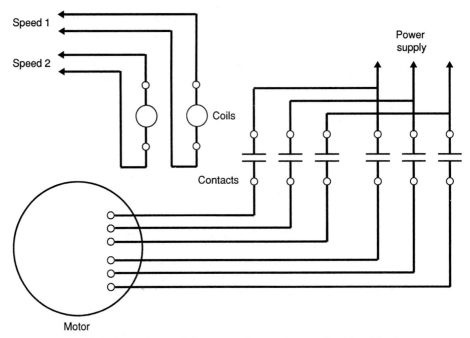

Figure 12-4 Stepped control of a pneumatic control system is achieved by the use of a stepped controller and a multistage heating or cooling system.

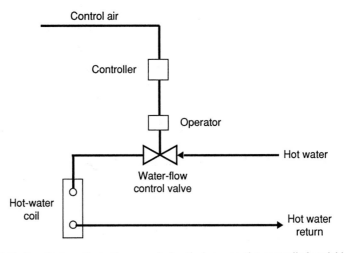

Figure 12-5 Sensitivity ratio is the numerical ratio between the controlled variable range and the control pressure range.

a certain variation in control output pressure per unit change in control variable. In a thermostat, the output pressure change must be set to coincide with the temperature change sensed. This is called the *sensitivity ratio* of the control.

Example A hot-water heating system with a direct-acting controller and a normally open valve operator is adjusted to have an operating range of 3 to 15 psi. The valve is half-open when the operator is at set point and the operating pressure is 9 psi. The system is adjusted for a sensitivity ratio of 10°F, to 2½ psi of operating pressure. If the temperature of the controlled variable goes up 1°F, indicating that less heat is needed, control pressure increases 2½ psi, to 8½ psi, and the valve closes slightly more than one-sixth of its travel. This reduces the flow of water through the coil and less heat is available in the space. This is the action desired. If the temperature in the space drops by 1°F, the control pressure goes down by 2½ psi, to 6½ psi. This causes the actuator to open slightly more than one-sixth of its travel, and more hot water flows through the coil to warm the space (Figure 12-6).

AIR-CONDITIONING SYSTEM CONTROLS

A pneumatic air-conditioning control system is one that controls the output of an air-conditioning system so that it will match the cooling load on a given building. The cooling load is the amount of control required to make up for heat gain to the building by external and internal effects. Three basic types of control systems are used: digital, stepped, and proportional.

Digital Air-Conditioning Control

In those systems where simple on–of or digital control of an air-conditioning system is desired, a pneumatic–electric switch is used. A pneumatic–electric switch is an electric switch that is actuated by a pneumatic signal. Control pressure from the controller in a system will trip a pneumatic–electric switch to turn an air-conditioning unit on or off.

Example A pneumatic control system operates an air-conditioning system in an on–off mode, through a pneumatic–electric switch (Figure 12-7). Control air pressure from a pneumatic air-conditioning controller operates the pneumatic–electric switch. The electric contacts in the switch are wired in series in the electrical control circuit to the air-conditioning equipment.

Stepped Air-Conditioning Control

Stepped control of an air-conditioning system requires that the air-conditioning unit used have a multispeed compressor motor, multiple compressors, or the capability of being unloaded. A cooling thermostat is used that produces a control output pressure that varies with any variation of the ambient temperature from set-point temperature. Control output pressure from the thermostat operates a pneumatic stepped controller in which a series of electric switches are turned on or off in sequence. The switches in the stepped controller are

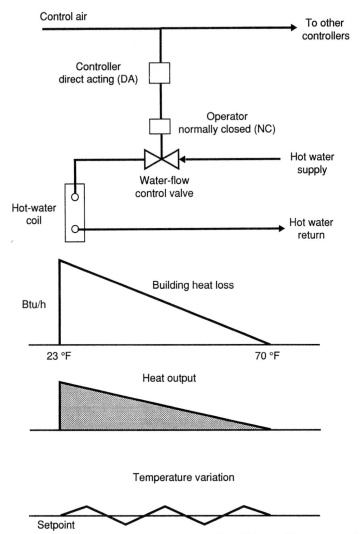

Figure 12-6 In a proportional control system the controlled variable conditions remain close to the set point.

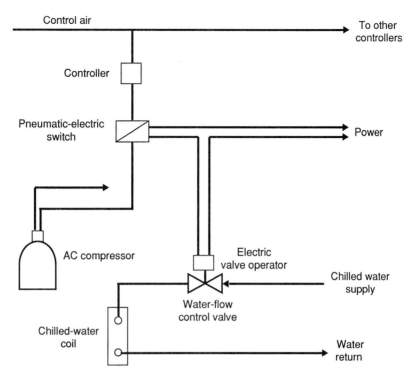

Figure 12-7   Digital control of an air-conditioning compressor using a pneumatic control system can be provided by using a pneumatic–electric control.

wired into electrical circuits to individual relays or contactors that operate the air-conditioning system in steps (Figure 12-8).

If the air conditioner has two compressors, a two-step controller is used and the compressors are controlled on the rise and fall of the temperature in the room in two steps. The same control would be used for a multispeed compressor. If the air conditioner incorporates unloading devices, they are controlled by the stepped controller.

Proportional Air-Conditioning Control

Proportional air-conditioning control is a natural function of a hydronic air-conditioning system, and a hydronic cooling system is nearly always used with a pneumatic control system. In a hydronic air-conditioning system, chilled water is used as the control medium, and a chilled water coil is used as the terminal device in the system (Figure 12-9). Flow of the chilled water through the coil is controlled by a modulating control valve. The valve is controlled by control pressure from a pneumatic controller.

> **Example** In a cooling-only air-conditioning system, the main control pressure may be either direct or indirect acting from 11 to 25 psi main pressure. The

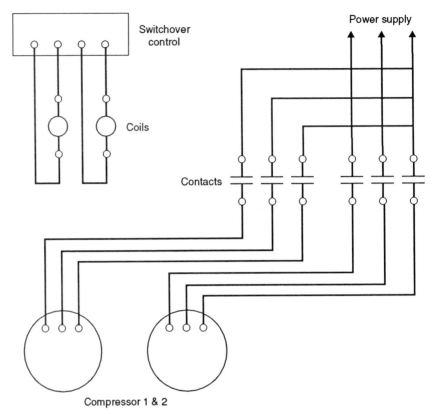

Figure 12-8 A typical application for stepped control is an air-conditioning system with dual compressors, controlled by a two-stage controller.

controller is connected by tubing or piping to a pneumatic valve operator. The valve operator may be either NO or NC. In a system with a direct-acting controller and an NC valve, the system is balanced with the valve at midposition when the controller is at set point. If the temperature in the space goes down, indicating a need for more cooling, the control pressure will also go down and the valve will open more. As the valve opens, more chilled water flows through the coil and the room is cooled.

Pneumatic controllers may control cooling only, or they may control both heating and cooling. Typically, to use a combination heating and cooling controller, a two-pressure air-supply system is used. The lower of two pressures is used for cooling and the higher for heating. A changeover switch is used to switch from one pressure to the other, and the controller automatically provides output pressure to match (Figure 12-10).

Example In a hydronic cooling system with a heating and cooling controller, the control pressure for cooling is 15 psi and for heating it is 20 psi. A system

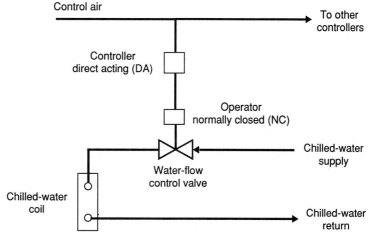

Control air

To other controllers

Controller direct acting (DA)

Operator normally closed (NC)

Chilled-water supply

Water-flow control valve

Chilled-water coil

Chilled-water return

Figure 12-9 A typical pneumatic control diagram shows the controller and actuators.

changeover control is used to switch from one pressure to the other (Figure 12-11). When the system is switched to cooling, 15 psi main pressure is routed to the controller. Output pressure from the controller varies from 15 psi maximum down to about 3 psi minimum. This output pressure is regulated by the cooling bimetallic element on the controller. The output pressure from the controller leads to a valve actuator. With 15 psi as a maximum main pressure, the controller functions as a cooling controller, and the valve is operated as a chilled water valve. Obviously, the water in the system must be controlled as chilled water for controlling and hot water for heating. When the system is used for heating, the control pressure is 20 psi. This is furnished by the system changeover control. The controller now functions as a heating thermostat and the output pressure from the controller operates the valve on the water coil as a heating valve.

If the temperature goes up in a space in which temperature control is desired, the control pressure from the pneumatic controller also goes up. At a control pressure relative to an increase of about 2°F, the pneumatic–electric switch will close and the air-conditioning system comes on. As the temperature in the space drops to about 2°F below the set point, the pneumatic–electric switch opens and the air-conditioning equipment goes off.

VENTILATION CONTROL

The control of ventilation air by a pneumatic system is basically the control of dampers. Typically, remote-bulb controllers are used to operate damper operators to provide mixed-air control, with monitoring of outdoor air, return air, and supply air. A remote-bulb controller has a sensing bulb that extends into the space being controlled (Figure 12-12). The remote bulb is connected to a pneumatic controller by a capillary tube. The relay provides variable output

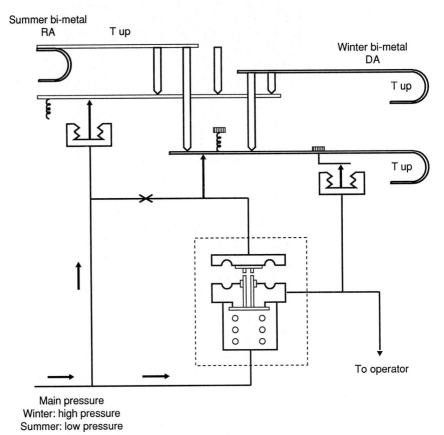

Summer bi-metal
RA   T up

Winter bi-metal
DA

T up

T up

To operator

Main pressure
Winter: high pressure
Summer: low pressure

Figure 12-10 Summer–winter or day–night control can be provided by using a special dual-pressure pneumatic controller.

output pressure, to coincide with the temperature of the control variable. The output pressure then controls the damper operators to position the dampers.

Example A damper control system that brings outside air into a building for cooling the building whenever the outside air is as cool or cooler than the supply air temperature desired is called an *economizer control system* (Figure 12-13). To use a pneumatic control system for an economizer control system, a duct/damper arrangement includes outside-air intake and dampers, return-air duct and dampers, and exhaust or pressure-relief connections.

The control arrangement is for a mixed-air controller to operate return- and outdoor-air dampers together, with the dampers arranged so that one set opens when the other closes. The exhaust-air damper should be slaved to the outside-air damper so that the two dampers open and close at the same time. Normal operation of the economizer system is when the outside air is less than the set-point temperature of the mixed-air controller, and the return air is about 80°F.

If the mixed-air controller is a direct-acting controller, and the damper operator is NC then, the mixed-air controller is set for about 55 to 60°F temper-

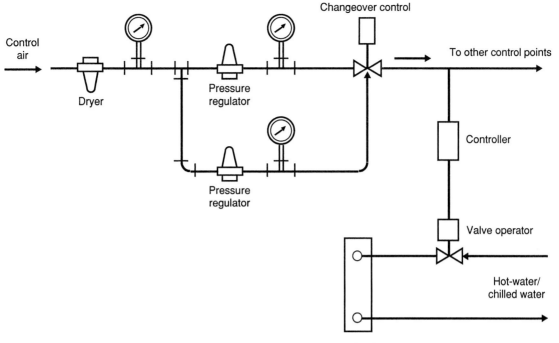

Figure 12-11 By using a two-pressure control system and a changeover valve, one controller can provide both heating and cooling control.

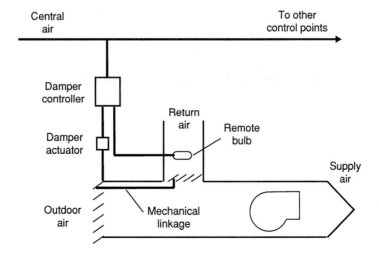

Figure 12-12 A remote-bulb controller is often used in pneumatic control systems on a damper package.

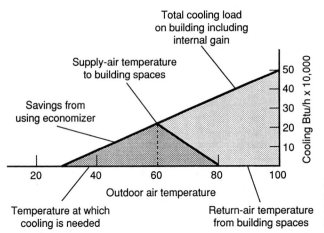

Total cooling load
on building including
internal gain

Supply-air temperature
to building spaces

Savings from
using economizer

50
40
30
20
10

Cooling Btu/h x 10,000

20   40   60   80   100

Outdoor air temperature

Temperature at which
cooling is needed

Return-air temperature
from building spaces

Figure 12-13 The savings realized by using an economizer control system can be graphed by plotting outdoor temperatures against building load.

ature. At set point the damper operator is set to be 50% open. If the temperature of the mixed air goes above the set point, the control pressure from the controller goes up. This causes the damper operator to open more than 50%, admits more outside air, and cools down the mixed-air temperature. If the temperature of the mixed air goes down, the control pressure from the mixed-air controller goes down, and the damper operator closes to less than 50% and less cool outside air enters.

SUMMARY

Proper application of pneumatic control systems requires knowledge of both the controls and the equipment to be controlled. Pneumatic controls provide three basic types of control: digital, stepped, and proportional, and all three types can be applied to HVAC system applications.

Some heating and air-conditioning equipment is designed to operate digitally, or simply in an on–off mode. Many combustion heating and single-compressor air-conditioning systems are in this category. These units are usually best controlled digitally. If the equipment can be operated in steps or stages, a stepped control system is often used. This includes heating equipment, in which heating output can be modulated or air-conditioning units with multiple compressors or with unloading capabilities. Hydronic heating and air-conditioning units lend themselves naturally to modulating control, and hence to the application of pneumatic controls.

QUESTIONS

12-1. *True or false:* It is possible to design a control system without any knowledge concerning HVAC equipment or the controls themselves.

12-2. Name three basically different types of heating control systems.

12-3. What is the outstanding feature of a digital control system?

12-4. To achieve stepped control with an HVAC system, the mechanical equipment must have what capability?

12-5. What does the term *sensitivity ratio* refer to in control system operation?

12-6. What is a pneumatic–electric switch, and how does it function?

12-7. If a pneumatic control system has one pneumatic–electric switch that operates the compressor in an air-conditioning system, which of the three types of control systems is it?

12-8. Stepped control of air-conditioning equipment requires that the equipment has certain capabilities. What are they?

12-9. How many switches would a stepped controller have to have to control an air conditioner with two compressors?

12-10. Why does a pneumatic control system lend itself so well to controlling a hydronic heating or air-conditioning system?

12-11. In a combination heating/cooling HVAC system using a pneumatic control system, what is changed in going from heating to cooling control?

12-12. When a direct-acting pneumatic thermostat is used to operate a valve on an air-conditioning coil, should the valve be NO or NC?

12-13. What is the name of a damper/control combination used to bring outside air into a building for cooling when it is cool enough outside to do the job?

12-14. What two sets of dampers are required for an economizer control package?

APPLICATION EXERCISES

12-1. On the accompanying chart show what the sensitivity ratio would be for each of the pressure/temperature combinations shown. Explain what sensitivity ratio is.

Sensitivity Ratio

Pressure Range	Temperature Range	Ratio
12 psi	35 deg °F	_____
15 psi	30 deg °F	_____
9 psi	25 deg °F	_____

12-2. Complete the drawing of the pneumatic control system on the following diagram of a pneumatic control system applied to a hydronic heating system. Name each of the controls and parts of the system by writing in the names on the tags on the diagram.

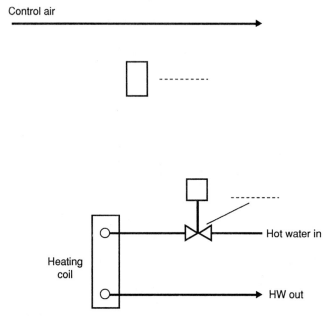

12-3. Mark on the scale above the piston operator shown in the accompanying diagram the position the piston would be at for pressures of 4 lbs, 7 lbs, and 12 lbs if the piston traveled its full length between 3 and 12 psi. What position should the piston be in when the pressure is at 9 psi?

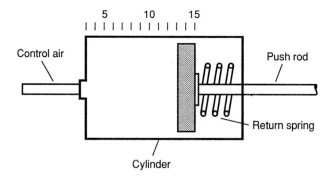

12-4. Complete the drawing of the pneumatic control system on the following diagram of a pneumatic control system applied to a hydronic cooling system. Name each of the controls and parts of the system by writing in the names on the tags on the diagram. Explain why the controller and actuator must be matched for proper performance.

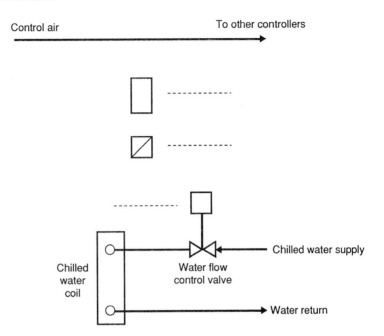

12-5. Complete the drawing of the pneumatic control system on the following diagram of a pneumatic control system applied to a hydronic heating and cooling system. Name each of the controls and parts of the system by writing in the names on the tags on the diagram.

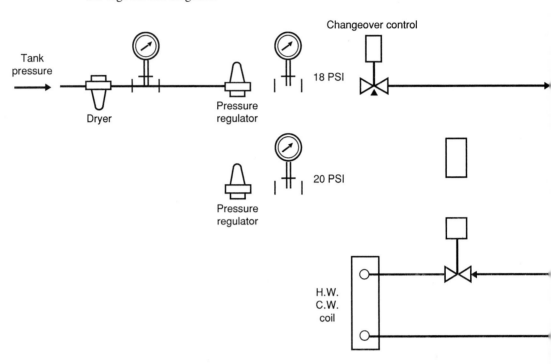

13

Electronic Control Systems

Electronic control systems are systems used to control the heating, ventilation, and air-conditioning equipment in a building with electronic components, to provide a comfortable climate in the building. An electronic control system is basically a proportional system. The components of electronic control system are generally solid-state devices. For some parts of the system, such as controllers, the devices, such as diodes, transistors, and other electronic components, are mounted on circuit boards and connected together to perform a certain control function. Because electronic controls provide excellent control in practical applications their use is growing rapidly.

CONTROL MEDIUM

The usual control medium for an electronic control system us varying-voltage direct current (dc). The control signal is either varying voltage or amperage, with variations that corresponds to differences between the controlled medium and a set point. The control signal is the output signal from a controller. In most cases the controller receives input from a sensor, and from that input generates an output signal that goes to an actuator or controlled device (Figure 13-1). The controller is an intermediate control device where the signal is

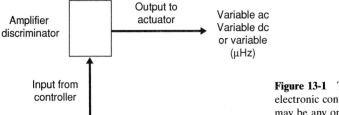

Amplifier
discriminator

Output to
actuator

Input from
controller

Variable ac
Variable dc
or variable
(μHz)

Figure 13-1 The output signal from an electronic control system main panel may be any one of several varying electronic values.

modified before it goes to an actuator. Variable-voltage systems use signals that vary from ±0 to 18 volts dc. Varying-amperage systems have a pulse width of 0 to 20 milliamperes. Communication is through regular control wire, shielded core cable, telephone lines, or by radio. All of the components in an electronic system pneumatic control system, including the actuators, must be matched so that they work properly with each other and so they each respond to the same control signals.

SYSTEM OPERATION

An electronic control system is inherently a proportional or modulating system. The basic parts of an electronic control system are sensors or transmitters, controllers, and controlled devices or actuators (Figure 13-2). A sensor performs the function of a thermostat, humidistat, pressurestat, or other sensing device. A transmitter is similar to a sensor but is usually a remote-bulb device and often includes an indicating function. In the case of a temperature control system the sensor normally contains a thermistor and a set-point adjuster. A *thermistor* is a solid-state device in which electrical resistance changes as the temperature changes. The controller serves the function of a central control

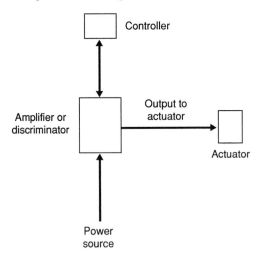

Controller

Amplifier or
discriminator

Output to
actuator

Actuator

Power
source

Figure 13-2 An electronic control system usually includes a central control panel called an amplifier or discriminatory, a controller, an actuator, and signal and feedback elements.

panel. It receives signals from the sensor and sends signals to the actuator (Figure 13-3).

Electronic Proportional Control

An electronic control system is basically a proportional or modulating system. Most electronic control sensing devices automatically follow varying, or analog signals. In this case, *analog* simply means changing. This means that the system can easily track changing control variables and produce an output signal that is proportional to the input.

In most proportional electronic control systems the sensor and set-point adjuster form one part of a bridge circuit, with the other part in the controller

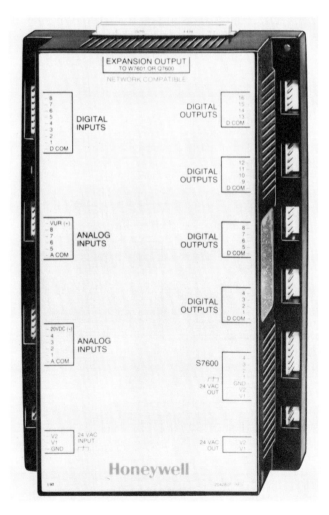

Figure 13-3 An electronic control center includes the components that produce a control signal, receive control medium data, and analyze those data to provide a signal capable of operating actuators.

(Figure 13-4). The bridge circuit may be in the controller itself, or it may be part of a printed circuit in some other part of the system. The basic control signal used may vary between ±0 and 18 volts dc or ±5 volts dc, or it may vary between 0 and 20 milliamperes pulse width, depending on the system used. In the controller this signal is converted to an output signal of 24 volts ac to the actuator.

Electronic Digital Control

While an electronic control system is inherently proportional or modulating, it can also be used as a digital system. In a digital electronic control system, an electronic signal, which may be varying, is used to trigger an electric or pneumatic control device (Figure 13-5). This is usually accomplished by having an electrical switch or pneumatic relay open or close when the input signal from the electronic sensor reaches a specific magnitude as it varies. Whereas the input signal from the sensor is analog, the output signal from the controller is digital.

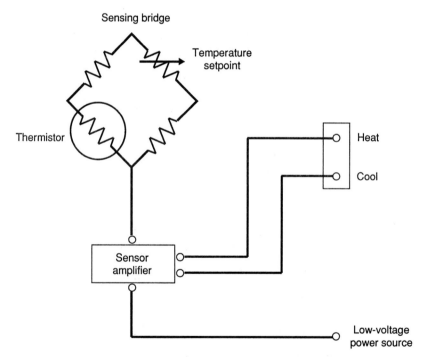

Figure 13-4 The bridge circuit is the heart of many electronic control devices.

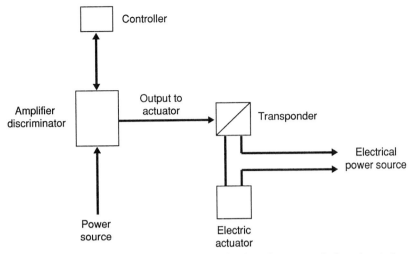

Figure 13-5 A transponder is a control device that allows control of an electrical circuit with an electronic signal input.

CONTROL DEVICES

The control devices in an electronic control system include the sensors, transmitters, controllers, and actuators or controlled devices.

Sensors

Sensors are the part in an electronic control system that senses the condition of a controlled variable and generates a signal corresponding to the condition (Figure 13-6). Sensors are available for sensing temperature, humidity, pres-

Figure 13-6 Sensors for electronic control systems are available for sensing temperature, humidity, pressure, and other control variables.

sure, and other variables related to building comfort and safety. An electronic device called a *thermistor* is often used as a sensor for a temperature. A thermistor is a crystal device in which the resistance to electrical flow changes in response to temperature changes. Other sensors are used for humidity in which electrical properties vary as humidity in the air varies.

Controllers

The *controller* in an electronic control system is the central control for the system (Figure 13-7). It contains the solid-state components that originate the signal that is sent to the sensor and receives a modified signal back from the sensor and then processes the signal to produce an output signal that goes to the actuator in the system. A number of different types of controllers are available. Some are used with either heating or cooling control; others are used with systems for heating and cooling control. Still others are used for multipurpose control, such as temperature and humidity combined.

Actuators

Several types of actuators, or controlled devices, are used in electronic control systems. Among them are those used for heating systems, air-conditioning

Figure 13-7 An electronic controller functions in the system by providing information to a discriminator relative to variations between setpoint and control variable.

systems, economizer damper controls, two-position and three-position damper control, digital control, and proportional control. In most cases, the control actuator is some type of electric switch, relay, or gear motor for valve or damper control.

Electrical switches, relays, and solenoid controls are used primarily as actuators when an electronic control system is used to control digital or on–off equipment. In this case control of the equipment is similar to the control of such equipment with an electric control system. Many electronic control actuators are gear reduction motors. They are used especially for two- and three-position damper motors. Some floating and proportional motors are also used. Many damper and valve actuators are piston or diaphragm type. They are used mostly for damper actuators, but in some cases as valve operators also. When an electronic control system is used to control equipment in a proportional mode, the operator used is usually motor-operated control (Figure 13-8). If a 24-volt ac damper or valve motor is used, the signal to the motor normally originates as output from a bridge circuit controller. An alternative choice is a hydraulic pump motor normally used in electronic control systems.

A hydraulic pump motor contains a small reservoir of hydraulic fluid, an electric hydraulic pump, and a three-way valve to control the flow of hydraulic fluid. The electric motor continually circulates the hydraulic fluid from the reservoir through an internal plumping circuit. A three-way diverting valve in the plumping circuit controls the flow of the fluid. When the valve is open, the fluid actuates a small motor or piston and provides valve operation. When the valve is closed, the fluid is diverted from the valve operator and simply recirculates within the plumbing circuit. The valve is controlled by electrical input from the control system sensor.

Transducers

A *transducer* is any device that converts one form of energy to another (Figure 13-9). In an electronic control system a transducer is one that changes an

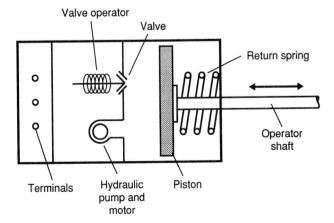

Figure 13-8 One actuator used in an electronic control system is a self-contained hydraulic motor actuator.

Valve operator

Valve

Return spring

Operator shaft

Terminals

Hydraulic pump and motor

Piston

Control Devices

electronic signal to an electric or pneumatic one. Transducers are used to interconnect two different types of control systems. In many electronic control systems a transducer is part of the controller. In systems where the control signal originates in the sensor package, the transducer is part of that package. Electric power is brought to the controller or sensor, and there it is converted to an electronic signal for electronic device operation.

There are one- and two-input electronic–pneumatic transducers (Figure 13-10). They receive either one or two electronic signals and produce either two proportional pneumatic outputs or one pneumatic and one electronic output. There are variable-input electronic–pneumatic transducers that receive a variable electronic input and produce a variable pneumatic output. There are also variable electronic–pneumatic transducers that are a combination of different transducers that produce multiple variable pneumatic outputs of different pressures.

To achieve electronic-to-electric transducing, an electronic–pneumatic transducer is often used with a pneumatic–electric switch. A pneumatic signal from the pneumatic–electric transducer controls a pneumatic–electric switch to provide electrical control. In a typical electronic–pneumatic transducer, an electronic signal of from 4 to 20 milliamperes dc is converted to a pneumatic signal of from 3 to 15 psi.

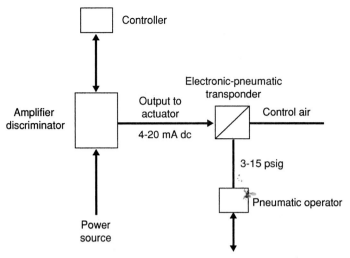

Figure 13-10 An electronic–pneumatic transducer functions as a relay in the pneumatic system to provide a control function in response to electronic signals.

SUMMARY

An electronic control system is one in which electronic control devices utilize a varying-power dc current as a signal, and variations in signal magnitude to indicate variations between control variable and set-point conditions. An electronic control system contains sensors or transmitters, a controller, and actuators. When used for digital control of heating or air-conditioning equipment an electronic control system usually utilizes devices called transducers to convert the electronic signals received from the controller to an electric or pneumatic signal. The actuators used in most of the systems are the same as would be used in an electric control system.

QUESTIONS

13-1. What control medium is used for an electronic control system?

13-2. What constitutes a control signal for an electronic control system?

13-3. What special function is performed by a sensor in an electronic control system?

13-4. *True or false:* When a bridge circuit is used as part of the control circuiting for an electronic control system, the sensor and set-point adjustment form one part of the bridge circuit.

13-5. *True or false:* An electronic control system can only be used as a proportional system.

13-6. Name three devices that are considered to be part of the controlled devices in the electronic control system.

13-7. Describe a typical sensor used for sensing temperature in an electronic control system.

13-8. Describe a typical sensor used for sensing humidity in an electronic control system.

13-9. What type of actuator is used as an operator in an electronic control system used for digital control?

13-10. Describe a hydraulic pump motor and explain how it works.

13-11. What is the function of a transducer?

13-12. How is an electronic–electric transducer similar to a pneumatic–electric switch?

APPLICATION EXERCISES

13-1. Draw a schematic representation of an electronic control system showing three major sections with the connecting electrical lines. Label the sections and the connecting lines.

13-2. On the attached diagram of the sensing section of an electronic control system name each of the controls and parts of the system by writing in their names on the tags on the diagram.

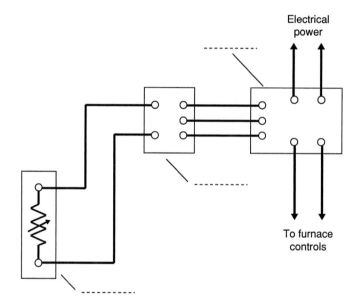

13-3. Name the control parts on the attached electronic wiring diagram by putting the names of the controls on the tags provided. Is this a digital, stepped, or proportional control?

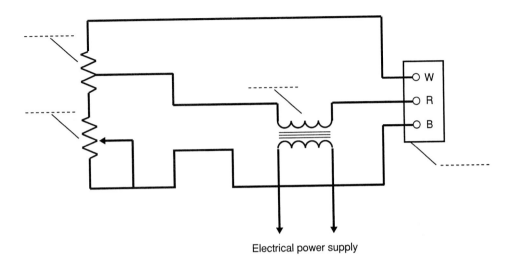

Electrical power supply

13-4. Name the two types of electronic sensors shown in the following diagrams, and describe how each works to change the temperature variations into electrical signals that can be used in a control system.

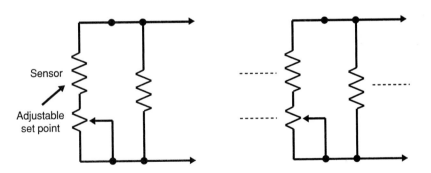

13-5. Complete the following wiring diagram to show how the electronic damper motor would be controlled in a typical ventilation control system.

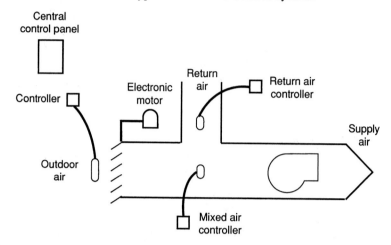

13-6. On the accompanying diagram of an electronic/hydraulic damper motor, show the sections that contain hydraulic oil. Cross hatch the section that is under high pressure, and use diagonal hatching to show the section that is under low pressure.

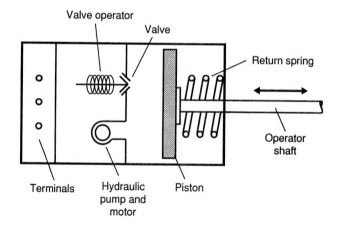

14

Electronic Control System Application

An electronic control system is one in which electronic controls are the primary controls used for operating the various parts and components of a heating, air-conditioning, or ventilation system. Low-voltage ac electricity is used as a signal medium. Control application is the way that controls are used to operate and control operation of the equipment used in the system.

ELECTRONIC HEATING SYSTEM CONTROLS

Controls used with heating systems are designed to control the operation of a heating unit so that the heat output of the unit will match the heat load on the building. The *heat load* is the total heat loss of the building, at the winter outside design conditions. Control of the operation of components or systems is dependent on the control of the individual parts in the component or system. The basic building block in a control system is the individual control operating a part or component of a system. Among these power controls are operating controls, safety controls, and combustion safety controls (Figure 14-1).

```
Power controls
    Overcurrent protection
    Disconnects
        Fusing
        Circuit breakers
    Transformers

Operating controls
    Thermostat
    Relays
        Magnetic
        Contractors
        Magnetic starters
        Starting relays
    Solenoids
    Blower controls

Safety controls
    Pressure switches
        High pressure
        Low pressure
        Oil pressure
    Motor overloads

Combustion safety
    Stack switch
    Flame surveillance
    Flame sensing
```

Figure 14-1 There are four categories of controls related to heating and air-conditioning system control with electronic controls.

Operating Controls

Operating controls are used primarily to turn the components and parts of a heating system off and on, and otherwise regulate the operation of the equipment in the system. Controls that sense changing conditions, such as temperature, humidity, or airflow, and operate the heating equipment in a system to compensate for those changes are examples of operating controls.

Thermostats. *Thermostats* are controls used to sense changes in temperature and to operate heating equipment to correct variations from a selected set point (Figure 14-2). In an electronic thermostat the sensing element most commonly used is a thermistor. A thermistor is a solid-state device in which resistance to electrical flow changes as the temperature changes. The thermostat receives a low-voltage dc signal from a control center or controller and then modifies the signal according to variations from set point. The signal from the thermostat goes back to the control center and there originates an ac electrical signal that is sent on to the actuator in the system.

Many heating systems are digitally operated. The heating unit goes on when the temperature at the thermostat goes below the set-point temperature by about 2°F and goes off when the temperature goes about 2°F above the set-point temperature. The on and off temperatures are above and below the set-point temperature by a given value called *offset*. In an electronic thermostat

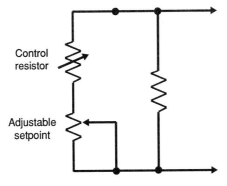

Figure 14-2 Basic electronic control consists of electronic circuitry and devices that measure variations between set-point and actual conditions and provide a signal relative to it.

the deviation from set point is measured in degrees Fahrenheit, but this is converted to output from the thermostat as variations in either voltage or milliamperes. At the controller the output from the thermostat is converted to a digital signal in low-voltage ac power. The output signal from the controller goes to the heating unit control relay.

In smaller heating units the thermostat often controls operation of the unit directly by operating a solenoid-operated gas valve on a gas furnace, a primary control on an oil burner, or a heating relay on an electric furnace (Figure 14-3). The thermostat is wired through a control panel to the heating unit actuator. In larger heating units, especially those combustion heating units with power burners, the thermostat operates a burner relay that operates the burner controls and the combustion safety circuits. In this case the thermostat

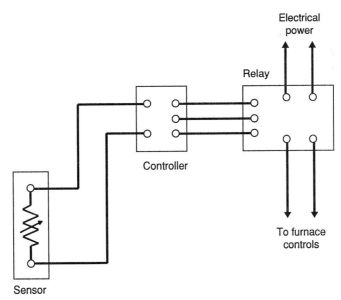

Figure 14-3 The signal from an electronic sensor is relayed to a controller, where it is modified as necessary and sent to an actuator.

Electronic Heating System Controls

leads are wired through a control panel to the control circuit to the relay coil. On a call for heating the relay is energized and the heating unit is turned on. When the call for heating ends, the relay coil is deenergized and the unit shuts down. If the unit has prepurge and postpurge cycles of the combustion air, for safe operation of the burner, the purge cycles are controlled by relays and controls combined with the combustion safety controls (Figure 14-4).

Motor controls.   Controls that operate fan motors in furnaces, called *blower controls* or *fan switches,* may be electronic devices. An electronic blower control can be used as a digital control simply to turn a motor on or off, but usually it is used to control the speed of the motor to control cubic-foot-per-minute output, in respect to some variable, such as temperature (Figure 14-5).

Example Some heating units use multispeed or variable-speed motors as blower motors. The motor is controlled by an electronic sensor in the supply air leaving the unit. The sensor functions as a variable resistance in one leg of a bridge circuit. The circuit sends an electric signal that varies with the air temperature to the motor controller. The cubic-foot-per-minute output of the blower varies as the heat output of the heating unit varies, to provide supply air at about the same temperature, regardless of heat-output variations.

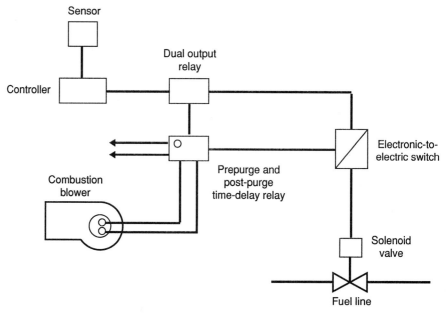

Figure 14-4   Prepurge and postpurge cycles of a combustion blower are controlled by time-delay relays in the line voltage circuit to the blower motor.

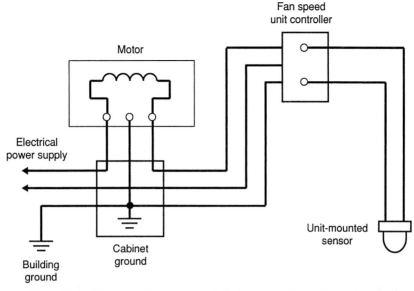

Figure 14-5 Motor speed control is relatively easy with an electronic control system.

Safety Controls

Safety controls are controls used to provide safe operation of heating equipment. Among the most common safety controls are limit controls of various types (Figure 14-6). Limit controls usually limit the temperature of the supply air of a heating unit to a predetermined limit. Limit controls on a typical furnace

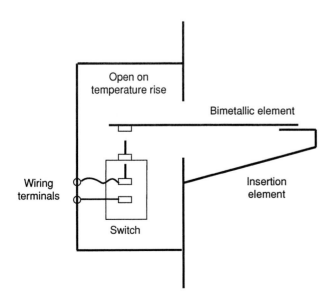

Figure 14-6 Many electronic devices provide automatic limit functions.

Electronic Heating System Controls

usually turn off the heating unit if the temperature of the air around the heat exchanger or in the supply plenum exceeds some limit. The usual limit for combustion furnaces is from 180 to 200°F.

Example An electronic limit control on a heating unit has a sensor with a thermistor in it. The sensor is in a probe located in the supply-air outlet of the furnace. A low-voltage dc signal is sent to the sensor by a control center. The voltage or amperage of the control outlet signal is controlled by the thermistor. Below the high limit temperature the signal to the control center, modified by the thermistor, signals that it is all right for the heating unit to continue to run. If the temperature sensed by the thermistor exceeds the limit setting, the signal through the thermistor calls for the heating unit to be shut off. The shutoff is usually made by opening an NC set of contacts in a relay. The relay is wired with the contacts in series in the control circuit to the heating unit.

In some cases a flow-proving switch may be used to determine if air or water is moving in a duct or piping system (Figure 14-7). The type of control is commonly called a *flow detector*. When duct heaters are used out in the branches of a duct system with a common blower on the trunk duct, it is usually necessary to prove the flow of air in the system before the duct heaters are energized. A flow detector is used for this purpose. Since most hydronic systems use a main water supply with terminal devices located along it, flow-proving switches may also be required on the pipe branches to prove water flow.

Example A heating unit with a centralized blower and a duct system with electric coils located in the branches for heating separate zones has flow-proving switches in the branch ducts ahead of the coils. The proving switches modify a dc electric signal that originates in a controller. If no air is flowing through the coils, the electronic sensor indicates this fact and sends a signal to a control center. The control center then sends an ac electric signal to an NC (normally closed) set of contacts in a relay. The contacts are in the control circuit from the thermostat that controls the heating coils. If air is flowing in the duct and through the heating coils, the sensor indicates that it is all right for the heating coils to be energized. If air is not flowing in the duct, the coils will not be energized.

Figure 14-7 Testing a fluid for flow conditions can easily be achieved with the proper electronic control devices.

Combustion Safety Controls

Combustion safety controls are controls used to make sure that a heating unit fires properly and safely on a call for heat, and also fires continuously during a combustion cycle. If a combustion heating unit does not fire when fuel and combustion air enters a combustion chamber, a dangerous condition is created. If the unit fires after the combustion chamber is filled with fuel and air, any spark or flame will set off an explosion. The equipment will surely be damaged and people in the vicinity may be endangered.

Electronic controls have long been used for combustion safety controls. Most electronic combustion safety controls are of the flame surveillance type. Flame surveillance controls monitor the flame in a combustion heating unit with some type of scanning device. The most common type of scanning device is a light-sensitive cell called a *photoconductive diode* (Figure 14-8). A photoconductive diode is an electronic device in which the resistance to an electric current varies in response to light. It is constructed with a thin film of light-sensitive material such as cadmium sulfide on a plate (Figure 14-9). The cell usually has a transparent protective lens that covers the film. Electrical leads are connected to the plate. These are connected to the control system.

Example In use, the photoconductive cell is placed where light from the pilot light or burner flame strikes the cell. When there is no flame and the cell is not

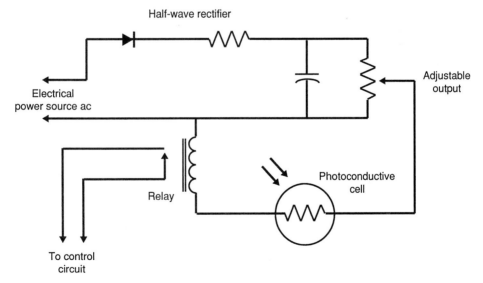

Figure 14-8 In an electronic combustion control system a photoconductive cell is used as a flame detection device.

Electronic Heating System Controls

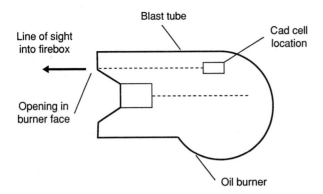

Line of sight into firebox

Opening in burner face

Blast tube

Cad cell location

Oil burner

Figure 14-9 A "cad" cell is one form of electronic device used in a combustion safety system.

receiving light, the resistance through the device is high. On a call for heat, when the fire is established and light strikes the cell, resistance through the cell is reduced. The photoconductive cell is connected by its electrical leads to an heating control center. The electrical signal from the cell, whether it indicates high or low resistance, is used to control an electrical circuit to the burner. If no light is indicated by the cell on a call for heat, the burner is shut down on safety. If light is indicated by the cell, the ignition process continues.

When used on gas-fired heating units, flame surveillance cells usually monitor a pilot flame. On a larger gas-fired heating unit, generally used for large commercial installation, a first-stage fire may act as a pilot for other stages. On oil-fired heating units the flame surveillance cell normally reads the fire in the combusion chamber directly.

ELECTRONIC AIR-CONDITIONING SYSTEM CONTROLS

Electronic controls are used extensively on air-conditioning systems. There are two basic categories of controls used: operating controls and safety controls.

Operating Controls

Operating controls are used primarily to turn the components and parts of an air-conditioning system off and on. Controls that sense changing conditions, such as a variation of ambient temperature from set-point temperature, and operate an air-conditioning system to compensate for those variations are good examples of operating controls. Operating controls also include humidistats, pressurestats, and motion-sensing controls.

Thermostats. *Thermostats* are controls used to sense changes in temperature and operate air-conditioning equipment to correct changes from a selected set point (Figure 14-10). In an electronic thermostat the sensing element most commonly used is the thermistor. A thermistor is a solid-state device in which resistance to electrical flow changes as the temperature changes. The

Figure 14-10 An electronic thermostat often contains a sensor and a microprocessor.

thermostat either receives a low-voltage dc signal from a control center and then modifies the signal according to variations from set point, or the signal is originated at the thermostat. In either case the signal from the thermostat goes to a control center, called a controller, and is then changed into an ac signal that is sent on to the actuator in the system.

In applications where smaller air conditioners are used, the air conditioners are usually operated digitally. This means that the air conditioner is turned on when the temperature at the thermostat goes about 2°F above the set-point temperature, and goes off when the temperature goes about 2°F below the set-point temperature. The 2°F above and below the set-point temperature is collectively called *control offset*. When an electronic thermostat is used, the deviation from set point is measured in degrees Fahrenheit but is converted to output from the thermostat in either voltage or milliamperes. At the controller the output from the thermostat is converted to an on–off signal in low-voltage ac power. This signal goes to a relay that turns the air conditioner on and off (Figure 14-11).

In most smaller air-conditioning systems the thermostat controls operation of the compressor through a control relay. The control signal from the central control panel goes to a relay with normally open (NO) contacts. On a call for cooling, the relay is energized and the contacts close. If the compressor motor draws less than about 15 amperes, the relay is used to operate it directly. If the motor draws more than 15 amperes, the relay is used as a pilot control to operate a motor contactor or starter.

In larger air-conditioning units, generally used in larger commercial applications, the thermostat often operates a liquid-line solenoid valve that controls the flow of liquid refrigerant into the evaporator coil(s) of the system. This is called a *pump-down control system* (Figure 14-12). In this type of control system, the low-pressure limit switch controls compressor operation. On a call for cooling the liquid-line solenoid is energized and the liquid-line valve opens. Refrigerant pressure in the low-pressure side of the air-conditioning system rises, and the low-pressure switch closes, allowing the compressor to come on.

Electronic Air-Conditioning System Controls

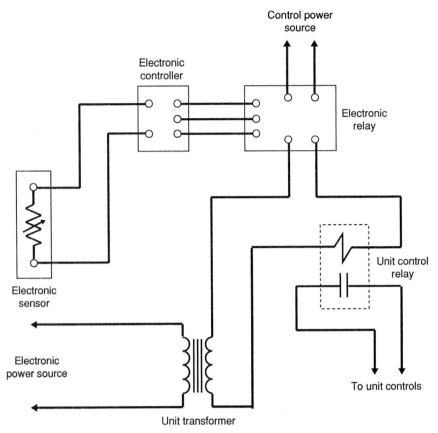

Figure 14-11 When electronic controls are interfaced with an electric system, a relay is generally used to make the connection.

When the thermostat is satisfied and no more cooling is needed, the liquid-line valve is energized, the valve closes, and the refrigerant pressure on the low-pressure side of the system drops low enough to cause the low-pressure switch to shut the unit down.

Motor controls. Electronic motor controls may be used to operate evaporator fan motors, condenser fan motors, compressors, or unloading devices in air-conditioning systems. Evaporator fans are the blowers that move the air across the evaporator coil and circulate it through a building. Condenser fans are those used on outdoor condenser coils for expelling heat from the system. Compressors are devices used for compressing the refrigerant in the system. Unloading devices are mechanical devices and arrangements of the refrigerant piping from the compressor to allow part-load operation of the air-conditioning system.

Evaporator fan motors are usually controlled by devices called *blower controls* or *relays* (Figure 14-13). These controls either bring on the evaporator

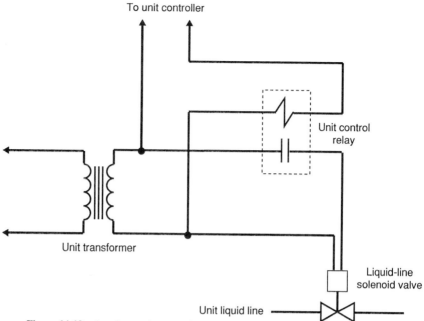

Figure 14-12 An electronic control system can be used to operate an air-conditioning unit in a pump-down mode.

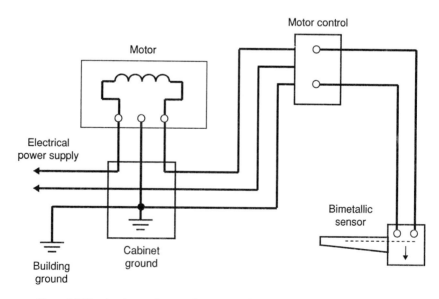

Figure 14-13 An electronic control system can be used with a variable-speed motor on the blower to provide constant air temperature.

motor when the system thermostat calls for cooling, or on call from a switch on the thermostat. The evaporator fan may be operated digitally simply to turn on and off. Or if a multispeed motor is used, the motor speed may be varied by the controls to provide variable airflow according to the temperature of the air. Electronic controls are especially suited to control the fan motor as a variable device.

When an air-conditioning system operates at cooler outdoor temperatures, the pressure in the refrigerant system can become low enough to cause frosting or freezing of the evaporator coil. To control the pressure, controls are used either to cycle the condenser fan motors off and on, or to vary the speed of the motors as the outdoor air temperature drops. This type of control is called low ambient control (Figure 14-14). For digital control, either an outdoor thermostat or a pressure switch on the discharge line of the refrigerant system cycles the condenser fan motors.

If a thermostat is used to control the condenser motors, it is operative only on a call for cooling. The thermostat turns the fan motors off if the outdoor temperature goes below a preset temperature and turns them back on if the temperature goes above the preset temperature. When a pressure switch on the discharge refrigerant line is used, the switch turns the motors off if the refrigerant pressure drops below a pressure corresponding to a freezing temperature for the refrigerant and turns the fans back on if the pressure rises above the same pressure (Figure 14-15).

Air-conditioning systems that operate with multispeed compressors, or with multiple compressors, require controls that change the speed of the compressor, or bring on the compressors, in response to the amount of cooling required. A proportional control system does this very well. A proportional

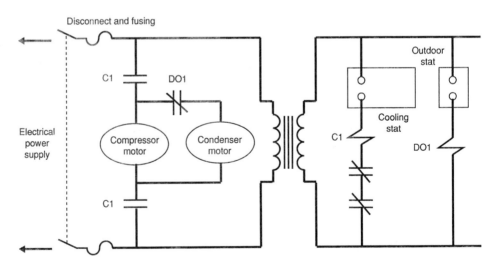

Figure 14-14 Low ambient control on an air-conditioning unit consists basically of cycling the condenser blower motors off and on. This may be done by sensing outdoor temperature.

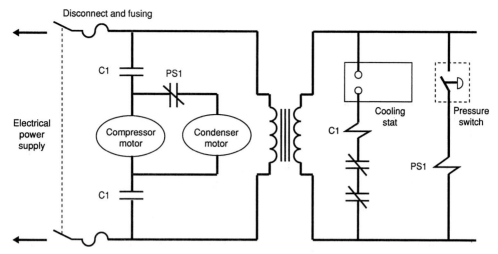

Figure 14-15 Low ambient control may be achieved by controlling the condenser fan motors in response to refrigerant pressure.

electronic control system uses a thermostat with a sensor as one resistor in a bridge circuit. Electrical output from the bridge circuit varies as the control variable varies. The output goes to a controller, where it is changed to an ac signal that energizes the motor operators in the system. If the motors are multispeed, the operators provide multispeed control. If multiple motors are used, the operator turns them off and on in sequence.

ELECTRONIC DAMPER SYSTEM CONTROLS

The control of dampers and damper systems to provide adequate ventilation to building spaces can be provided by the use of electronic controls (Figure 14-16). In addition, damper systems can be used to control the introduction of outdoor air to provide cooling when the outdoor temperature is lower than the

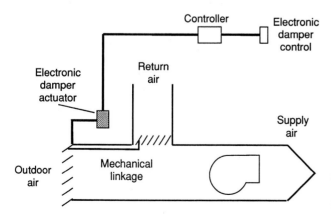

Figure 14-16 Damper motor control is usually a function of air temperature at some point in the controlled medium.

building temperature desired. Control of ventilation air dampers usually consists of opening the dampers to admit a predetermined amount of air during the periods of time that a building is occupied, and closing them during unoccupied times. To achieve this control, a manual switch or an automatic control that will indicate the presence of people in the spaces ventilated is used to send a control signal to a central controller. In some applications the system is used for an entire building, but in others it is used in specific spaces where ventilation is especially important, such as rest rooms or dressing rooms.

Example A typical application in a rest room would have a motion sensor located in the entrance to the room. A sensor would sense the presence of people in the room and send a signal to the controller. The controller interprets the signal and then sends a signal to the damper actuator(s) to open the dampers. The sensor in this case may be a motion sensor that is activated by the presence of a person in the room, or it may be activated by the opening of a door.

When a control system is used to control a set of dampers to provide ventilation air for cooling a building, it is called an *economizer cycle control system* (Figure 14-17). A mixed-air duct system is used that has a damper on a duct from an outdoor air inlet and another damper on the return air ductwork for the distribution system. The two dampers, outdoor air and return air, are linked together so that when one opens, the other closes. An exhaust damper or a relief air damper should be installed in the return air duct and should be linked to the return-air damper so that it will open to exhaust air from the building when the return-air damper closes.

To control the economizer damper package, a mixed-air control sensor is located in the mixed-air plenum. The control contains a thermistor as a sensor.

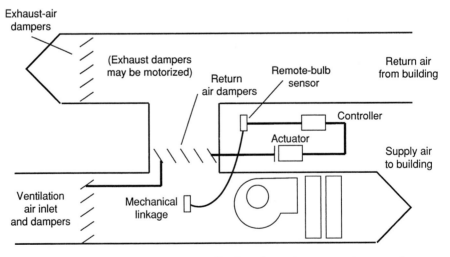

Figure 14-17 Electronic controls are easily adapted to use in an economizer control package.

This control is usually a remote-bulb sensor with an averaging bulb. An averaging bulb sensor is a sensor with a long bulb that is placed in the airstream in such a way as to provide a reading of the average conditions in the duct. The mixed-air control is set to maintain the same temperature as the required supply-air temperature for the applications.

Example In a typical application the mixed-air temperature controller is set to provide air of about 60 to 65°F. At the same time the return-air temperature from a typical cooling application is about 80°F. If the temperature at the mixed-air controller varies up or down from the set-point temperature the voltage reading from the sensor to the control center indicates a need for more or less outdoor air. As the outdoor-air damper opens, the return-air damper closes. As long as the outdoor air is cooler than the set-point temperature the dampers will open and close to maintain the set-point temperature. If the outdoor temperature goes above the set-point temperature, the outdoor-air dampers will start to close and more return air will be used. When an economizer damper and control system is used, the system fan is run continuously. During unoccupied times the damper control system can be programmed to return the dampers to minimum setting and have the fan motor cycle as needed to maintain the desired indoor conditions.

SUMMARY

An electronic control system is one that uses electronic control components to operate HVAC mechanical systems. An electronic control system uses low-voltage dc electrical power for a control medium, and voltage or amperage variations as control signals. An electronic control system operates naturally as a proportional system, but it can be used to operate equipment digitally also.

Electronic heating controls include operating controls, safety controls, and combustion safety controls. When used to operate an air-conditioning system there are only operating controls and safety controls. Electronic controls are especially well adapted for operating a damper package as an economizer system. When combined with modulating damper motors or valve motors, natural proportional control is achieved.

QUESTIONS

14-1. What is the basic requirement for a heating control system relative to heat output?

14-2. What is the main function of the operating controls of a system?

14-3. Match the term in the left column with the phrase that best matches it in the right column by placing the letter that precedes the description in the space provided preceding the term.

A. _____ Thermostat	a. measures humidity
B. _____ Thermistor	b. central control panel
C. _____ Humidistat	c. sensing element
D. _____ Controller	d. measures temperature

14-4. What does the term *offset* mean in relation to control operation?

14-5. What are two names for controls that operate blower motors?

14-6. What is the main function of the safety controls used in a heating unit?

14-7. Why is it usually necessary to check for airflow through a duct heater before the heater is allowed to come on?

14-8. *True or false:* It is only necessary to check initial startup ignition with a combustion safety control.

14-9. Describe the main feature of a flame surveillance combustion safety control.

14-10. Name the basic categories of electronic controls used in air-conditioning control systems.

14-11. *True or false:* With an electronic thermostat, deviation in temperature is converted to output in voltage or amperage.

14-12. What control actually turns the compressor on and off when a liquid pumpdown control system is used on an air-conditioning unit?

14-13. *True or false:* Unloading is the control of an air-conditioning system in such a way that cooling output more closely matches cooling load.

14-14. What is the control function called *low ambient control* used for?

14-15. *True or false:* A bridge circuit is used only in digital control systems.

14-16. *True or false:* An economizer damper system has two basic dampers, one on the return-air duct and another on the outdoor-air duct.

14-17. *True or false:* On an economizer damper control system a mixed-air temperature controller is mounted inside the building by the thermostat.

APPLICATION EXERCISES

14-1. Complete the following diagram of an electronic control system used on a gas furnace by drawing the control wiring necessary.

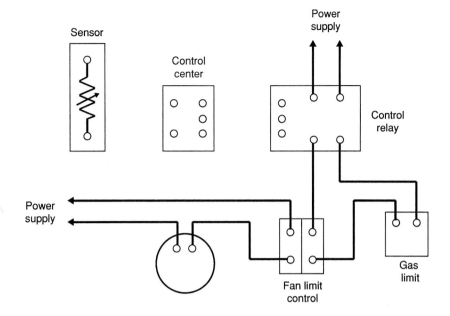

14-2. Complete the following diagram of an electronic control system used to control an electric motor by drawing the control wiring necessary.

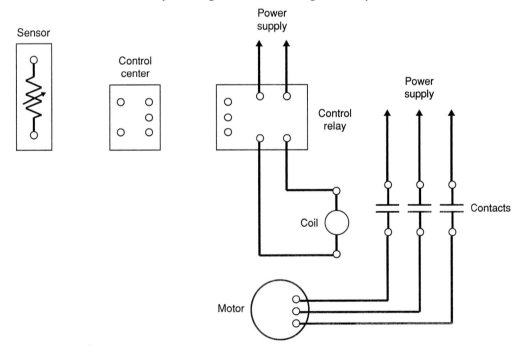

14-3. Complete the following diagram of an electronic control system used with a sail switch to prove air flow before an electric duct heater can come on by drawing the control wiring necessary.

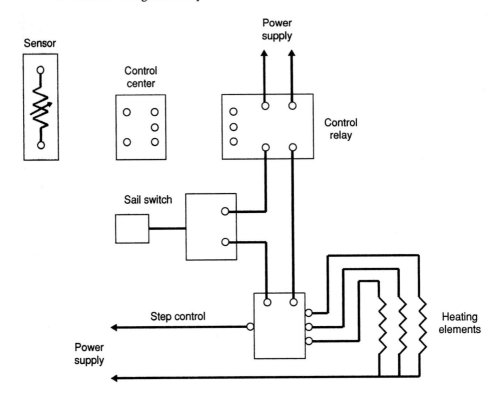

Electronic Control System Application Chap. 14

14-4. Indicate on the attached diagram those parts of the control system that are directly part of the combustion safety control by drawing a colored line under them. What specific condition related to combustion does this part of the control system protect against?

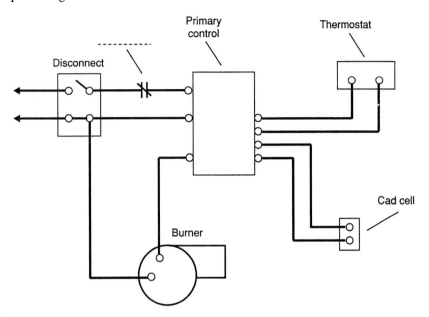

14-5. Complete the following wiring diagram to show how the electronic control system is used to control the operation of a typical air conditioning unit. Is this particular air conditioning unit controlled directly by the thermostat or is it a pump-down control system?

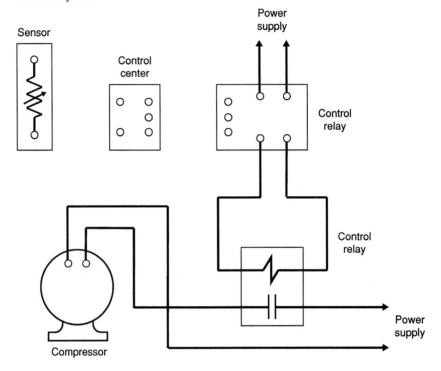

Electronic Control System Application   Chap. 14

15

Automated and Programmed Controls

An automated control is one that operates a piece of HVAC equipment automatically in response to commands that are stored electronically in the control. The control may be a thermostat, hudidistat, or a flow meter. Data relative to desired conditions, such as temperature in case of a thermostat, are entered into the electronic storage unit in the control through a keyboard or pushbutton arrangement. A sensor in a control compares actual conditions with those in storage and originates output signals in response to differences between the controlled variable and stored condition (Figure 15-1).

Digital, stepped, and proportional controls all require manual setting, and in some cases, manual manipulation for operation. Automated controls only require input of data in reference to desired conditions. Output data for actually controlling a system are provided by the automated control system. Programmed controls are similar to automated controls, with the exception that the commands to operate the mechanical equipment are stored in a central control panel. Both sensors and actuators are located remote from the central panel. Communication between the various parts of the system are by low-voltage ac or dc electric circuits.

Analog Control

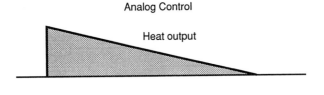

Heat output

Temperature variation

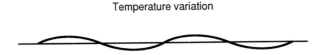

Figure 15-1 Analog control is based on tracing changes as they occur between a set point and a control point.

AUTOMATION

An automated control system is operated by a *microprocessor* (Figure 15-2). A microprocessor is a self-contained minicomputer that has a central processing unit (CPU) and input/output (I/O) capabilities. The central processing unit is made up of one or more printed circuit boards with electronic control circuits and devices that have the capability of processing input data in a logical way and providing output data in response to the input. Basic input data are in the form of commands that are input by keyboard entry and are stored for reference. A second source of input data is in the form of analog or digital signals received from sensors and controllers in the system. The output signals are in the form of analog or digital signals sent from the central processing unit to control devices in the system. Analog signals are continuous in relation to changing control data. Digital data are in the form of on–off signals.

Data are input at the keyboard in digital, or numerical, form. The microprocessor in the unit converts these data into electronic signals that are processed by logic circuits in the unit, and they are then converted to output signals. Typically, a microprocessor is very small. The microprocessor used in automated control is usually part of the control.

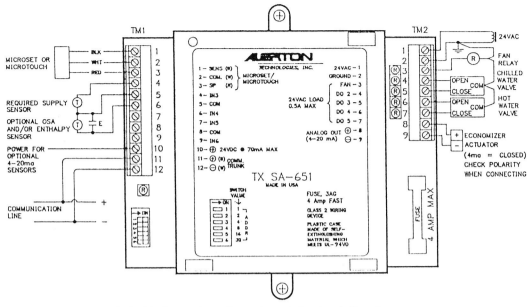

Figure 15-2 The heart of any electronic control system is the microprocessor.

PROGRAMMABLE CONTROLS

A *programmable control* contains a sensor, or sensors, a microprocessor, and some type of input capability. The input is usually by pushbutton and allows the input of set points and other data relative to the control desired. The control may be a programmable thermostat, humidistat, or pressure control. In a programmable thermostat, a clock in the control monitors time passage and allows programming for action at regular time intervals or at some future time (Figure 15-3). This makes it possible to input data relative to temperature setback for certain hours. Output from the control is usually limited to the opening or closing of a set of electrical contacts. HVAC equipment is then controlled by

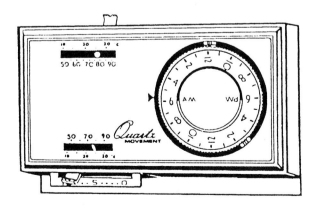

Figure 15-3 A programmable thermostat allows an operator to dial in conditions desired.

Programmable Controls

a low-voltage ac signal originating at the equipment but controlled by a switch in the programmable control.

> **Example** A programmable control system is used to monitor temperature in a building and control heating and/or air-conditioning equipment to condition the building. The programmable control is located where a thermostat would normally be placed. The control is "set" for control of either heating or cooling, the heating and cooling temperatures desired in the building, and for time-of-night setback and how much setback is desired. The clock in the control keeps track of time, and the sensor in the control monitors the ambient temperature.

The ability of a programmable control to output signals based on logical decisions makes it a good system to use for operating equipment in modes such as night setback or morning warm-up (Figure 15-4). A program can be input to cycle the equipment on and off at specific times, or only when the indoor and outdoor air temperatures indicate the proper times for startup or shutdown.

PROGRAMMABLE SYSTEMS

A *programmable system* is one in which a central control panel contains the electronic devices, including a microprocessor, for interpreting input data and initiating output data (Figure 15-5). Input data originate in sensors for temperature, humidity, pressure, or air motion, and may be in either digital or analog form. Output data are in either digital or analog form. The central control panel usually has a keyboard of some sort incorporated with it with which it can be programmed for various control functions. The main difference between a programmable control and a programmable system is that the system can process a number of inputs and outputs at the same time and operate

Input		Output
A	B	C
No	No	No
No	Yes	No
Yes	No	No
Yes	Yes	Yes

Figure 15-4 Logic gates are electronic devices or circuits that produce an output signal only in response to specific input signals.

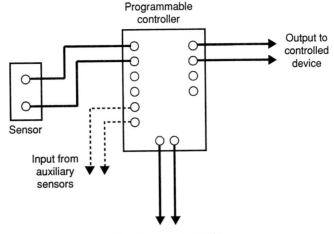

Figure 15-5 A programmable control system is one in which output on given control circuits is controlled by input from specific control points, according to a program entered into the system.

several pieces of equipment at the same time. A programmable control normally controls one piece of equipment at a time.

The control signal in a programmable system originates in a central control panel as a low-voltage dc signal. This signal is sent to a sensor used in each control circuit. The signal is modified in the sensor according to controlled variable conditions (Figure 15-6). For instance, in a temperature control system, the sensor is a thermistor in which electrical resistance varies in response to temperature. The thermistor may be direct acting or indirect acting. In a direct-acting thermistor the resistance goes up as the temperature goes up, and the resistance goes down as temperature goes down. If the resistance of the thermistor goes up, the increased voltage drop through the thermistor will cause a voltage drop in the control circuit, which is the control signal back to the control panel. This control signal back to the panel is the input signal.

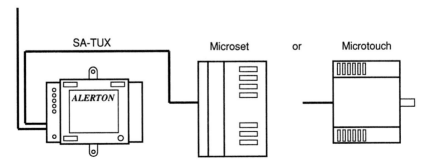

Figure 15-6 Most programmable control systems contain relays that control external low-voltage circuits to operate actuators in a mechanical system.

Programmable Systems

The input signal from each circuit in a programmable control system is processed by the microprocessor in the panel, and an output signal is generated to operate a relay in the panel (Figure 15-6). One terminal of the low-voltage side of this relay is connected to one terminal from a control transformer. The other terminal from the control transformer goes to the controlled device used in the system and then back to the other side of the relay in the control panel.

Programmable controls and control systems are used to control equipment digitally, but are designed so that the microprocessor samples the input at frequent intervals and the results are very nearly proportional. They are used with electric, pneumatic, and electronic controls. In an electrical digital system the output from a programmable control actuates solenoids, relays, contactors, or magnetic starters, depending on the parts being controlled. In a pneumatic system the output signal normally goes through an electric–pneumatic control to operate the equipment. In an electronic system the output signal goes to electric or electronic controllers or actuators to operate the equipment.

One of the advantages of using a programmable control system is that the system can be communicated with through normal telephone lines (Figure 15-7). A telephone device called a *modem* is used with the control panel to communicate with the microprocessor in a programmable system. Through use of this device a readout of operating conditions of the heating, air-conditioning,

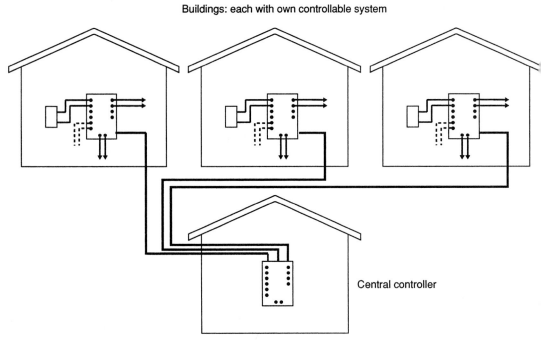

Buildings: each with own controllable system

Central controller

Figure 15-7 Most programmable control systems can be communicated with by telephone from a central point for monitoring or adjusting.

or ventilating equipment in a building can be acquired, or the controller can be reprogrammed from the controller. This type of operation allows the monitoring and control of HVAC equipment in any number of buildings to take place from one central location.

SUMMARY

Automated control systems combine the use of electronic controls with a computerized processor to make it possible to program controls to perform the process of controlling HVAC equipment, but in a preprogrammed manner.

A programmed control is one in which a control program can be input so that the control will automatically operate the equipment to which it is connected. In addition to set points for heating and air conditioning, seasonal and night-setback data can be input. A clock in a programmed control makes it possible to input data changes to occur at different time intervals.

A programmable system is one in which input can be programmed but for many different input/output circuits at the same time. It can also be programmed to take input in either digital or analog form. The input originates in signals from sensors that can be used to sense temperature, humidity, motion, or other thermodynamic or electrical characteristics. A programmable system can be communicated with by telephone through a modem, to allow monitoring or changing of set points from a central point for many different systems.

QUESTIONS

15-1. What is the device called that operates an automated control system?

15-2. The central processing unit used in an automated control system is capable of processing _____data in a logical way, and providing _____ data in response to the input.

15-3. *True or false:* Output signals in an automated control system are used to operate or control the operation of actuators of mechanical systems.

15-4. In what way is a programmable thermostat like a conventional thermostat?

15-5. What makes a programmable control an especially good control to use for night-setback control systems?

15-6. What makes a programmable system different from a programmable control?

15-7. *True or false:* The control medium used in a programmable control system originates in the central control panel.

15-8. *True or false:* The control signal in a programmable control system is the product of a sensor.

15-9. It is possible to communicate with a programmable control system by telephone. Why is this such an advantage?

APPLICATION EXERCISES

15-1. Identify the parts of the automated controller shown in the accompanying diagram by writing in their names on the tags provided, and answer the following questions:

      1. Is it necessary to have a separate power supply on the circuits connected to the output terminals?

      2. What part of the control center has a logic producing capacity?

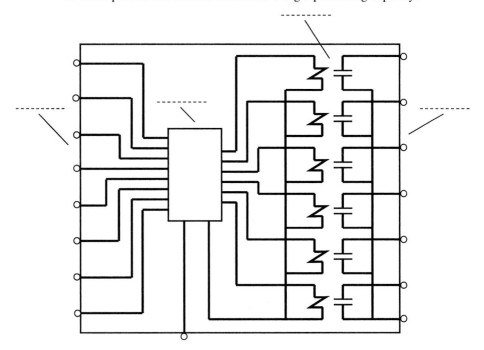

15-2. Draw in the lines of communication between a central control panel and separate buildings with control systems on the accompanying diagram. Answer the following questions:

   1. What is the name of the device used to make communication possible between control systems in distant buildings and one central control panel?

   2. What physical lines of communication are used between a central control system and separate control systems?

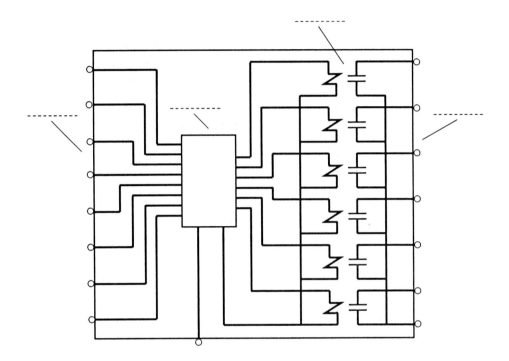

Buildings: each with own controllable system

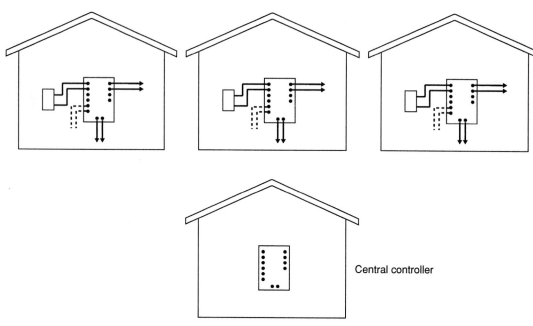

Central controller

15-3. Write in three possible types of output signals at the point where they occur on the accompanying diagram. Answer the following:

    1. Is it necessary to select the actuators to match the signal? Why?

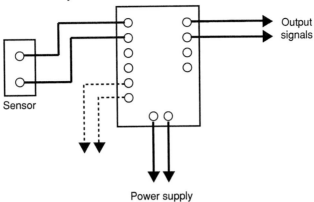

15-4. The following graph shows how a digital sensor follows changing conditions. How does a CPU interpret this information to produce proportional output? Draw a line on the graph to indicate that output.

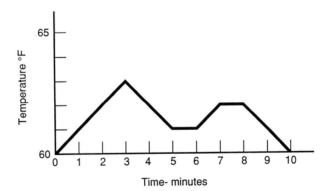

16

Direct Digital Control

Direct digital control (DDC) is the name given to a computerized electronic control system. Using a computer with a control system makes it possible to use a stored program for controlling the operation of the heating, air-conditioning, and ventilating equipment in a building. It also opens many other avenues as far as programming operation with other systems in the building, such as light control and energy-saving systems. An operating program that is stored on computer disks or tape contains information as to how the mechanical systems in the building should function. The use of a computerized control system allows scheduling of operating procedures based on time of day, day of week, and even annually. Interruptions to regular scheduling due to holidays or vacations can even be programmed in.

DDC SYSTEM

A computerized control system is comprised of a computer used for storing and running a program and for producing a visual image of system operation; and a central control panel, usually containing a microprocessor and relays for sending output signals (Figure 16-1). In addition, there are sensors that input

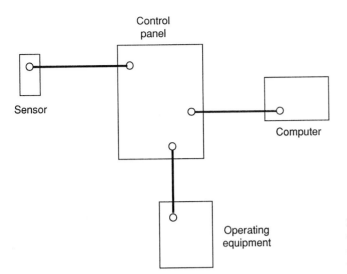

Control
panel

Sensor

Computer

Operating
equipment

Figure 16-1 A DDC control system includes sensors, control panel, computer, and actuators in the operating equipment.

data to the control panel, and actuators or operators that control the operation of equipment.

Computer

A computer used in a DDC system is a combination of electronic and mechanical equipment. A basic computer system includes a central processing unit, which includes a microprocessor; input devices such as a keyboard and a written program on disks or tapes; devices called disk drives for running the disks or tapes, to make the stored program available and a monitoring device, such as a screen (Figures 16-2 and 16-3). A printer is often also used as an

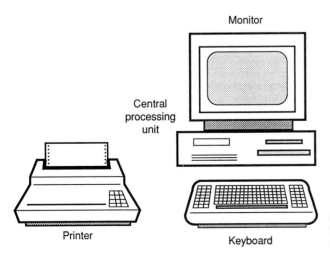

Monitor

Central
processing
unit

Printer

Keyboard

Figure 16-2 A typical computer system includes a central processing unit, input devices, and output devices.

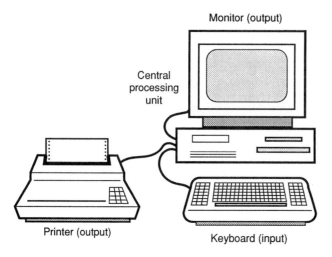

Monitor (output)

Central
processing
unit

Printer (output)

Keyboard (input)

Figure 16-3 The input devices in a computer system may be a keyboard, disk drives, or a pointing device. Output devices usually include a monitor, a printer, and disk drives.

output device in addition to the screen, for making copies of output information relative to equipment operation. Input to the computer is from sensors located at different control points in a building, from the program stored on disks or tapes, and can be by data input through the keyboard.

Use of a computer in a control system makes it possible to preprogram the operation of equipment, save the program on a tape or diskettes, and have the program run automatically. The program is written in one of the computer languages. Computer languages are commands written in a format that the computer can understand. The microprocessor in the computer makes logical decisions based on the commands, and outputs digital signals in response to the input.

A computer program written for a controllable system may include instructions to turn equipment on and off at certain intervals. It can also be programmed to turn the equipment on or off on in response to input signals such as temperature, humidity, pressure, or one of many other variables. The program can be written so that inputs from more than one control point are analyzed and output signals are related to the combined information from those inputs.

Central Control Panel

The central control panel in a computerized control system contains the microprocessor that receives the input data from the sensors, processes that input through various logic circuits, outputs data to a computer, receives data back from the computer, and sends output signals to the operating control devices (Figure 16-4). The microprocessor in the central control panel operates on the same principles as the microprocessor in a computer. Data received in digital form are processed through logic circuits to provide an output signal.

Figure 16-4   The central control panel contains a microprocessor and other circuitry to deal with input and output signals.

Microprocessor Logic

The function of a microprocessor as a decision-making instrument depends on its ability to apply the laws of logic to data input to it (Figure 16-5). Initially, input data are either in number or letter form. The input is converted by the microprocessor to combinations of electronic signals. The electronic signals are basically combinations of on-and-off electrical pulses. Each combination of pulses has a specific meaning. This is in accordance with a type of algebra called *Boolean algebra,* named after the mathematician who developed the system. For instance, in a system where two signals only are used, they would be an on signal and an off signal. If neither input signal is on, the output signal is not on either. If one input signal is on and the other off, the output is on. If both input signals are on, the output signal is on. By applying this type of logic, a microprocessor can convert digital input to practical output signals for operating equipment. Because of the short time spans between the input signals, the output signals are almost continuous in nature.

Input data.   Input data to a computer in a control system are normally digital in form, but the signals are monitored by the central processing unit at short intervals. Any change in the control variable becomes almost immediately

DDC System   **263**

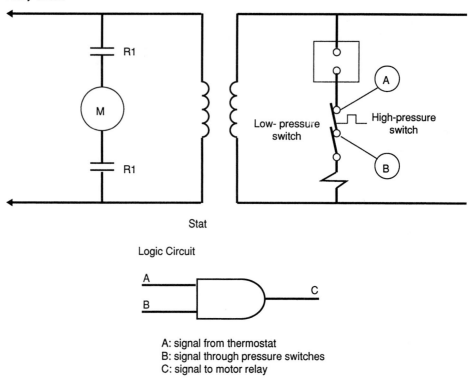

Relay Circuit

R1

M

R1

Stat

Low- pressure switch

High-pressure switch

A

B

Logic Circuit

A

B

C

A: signal from thermostat
B: signal through pressure switches
C: signal to motor relay

Figure 16-5 The operation of a DDC control system depends on a computer's ability to apply logic to input signals and provide output in response.

apparent. The intervals at which the variable is tested are usually less than 1 minute. Output signals that are originated in response to input data also change constantly to reflect input changes.

Output data. The output signals from a computerized control system are usually in the form of digital output from the control center. The output is sent to the control devices that operate various pieces of equipment in the system. Several output circuits can be utilized at one time by a typical DDC system. Each output circuit would normally be controlled in input from a particular input circuit.

DDC SYSTEM APPLICATION

DDC systems are usually used for control of the heating, air-conditioning, and ventilation systems in large buildings. The technology involved is practical for a building of any size, but the sophistication of control and the cost of the system both make it more practical for use in large buildings. A typical system includes sensors in various rooms or spaces in a building for sensing temperature, humidity, air pressure, or other condition relative to the operation of a comfort control system. Other sensors may be located outside the building, to provide outdoor reset for inside controls; on blower bearings to determine proper operation; and on electric motors to monitor electric power usage (Figure 16-6). The sensors are connected to a central control panel that processes signals from the sensors. Data related to the sensor readings are sent to a computer for comparison with parameters set up in the program stored on disks or tape. The data are usually also stored on disk to provide an operational log

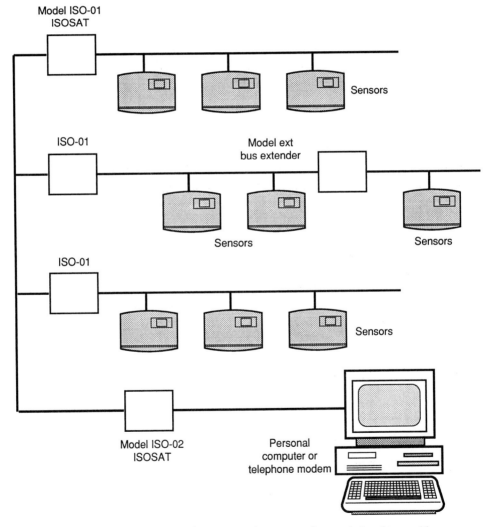

Figure 16-6 Input to a DDC system requires sensors that are designed to provide an input signal related to the particular variable involved.

of the system. Deviations from set points or other parameters in the stored program initiate a signal back to the central control panel, where another signal is originated and sent out to the operating equipment to correct the problem causing the deviation.

Sensors

The sensors used in a DDC system may be either digital or analog (Figure 16-7). Digital sensors are used to indicate variations above or below a set point by some predetermined value; usually about 2°F in temperature, or about 5% relative humidity. Analog sensors are used to specific information concerning a control point at any given time. Analog sensors are used to track variables that change continuously, such as outdoor conditions or electric energy use.

Actuators

Actuators used in DDC systems are usually the same as those used in an electric or electronic control system. In most cases the output signal from the central control panel to the actuator is a digital signal that can be used to operate a typical electric actuator such as a relay, solenoid control, or other electric operator. If a DDC control system is interfaced with a pneumatic control system, a pneumatic–electric switch is used for the interconnection between the two systems.

SUMMARY

A direct digital control system (DDC) is one in which electronic controls are used in conjunction with a computer to control the mechanical equipment automatically in an HVAC system. The computer is used to control the operation of the equipment by use of a program that is stored on a disk or tape and to monitor that operation.

A DDC control system includes sensors that are used to sense existing conditions related to any of the variables related to a total comfort system within a building. The sensors send signals relating to conditions to a central control panel. The central control panel contains a microprocessor that analyzes the signals and sends them on to the computer. At the computer the signals from the CPU are compared with set-point data found on the stored program, modified by the program if necessary, and sent back to the CPU and then on to actuators in the mechanical equipment.

The sensors used in a DDC system must be matched with the system but can be used to sense temperature, humidity, flow, energy use, pressure, and other characteristics relative to building operation. The system itself has the

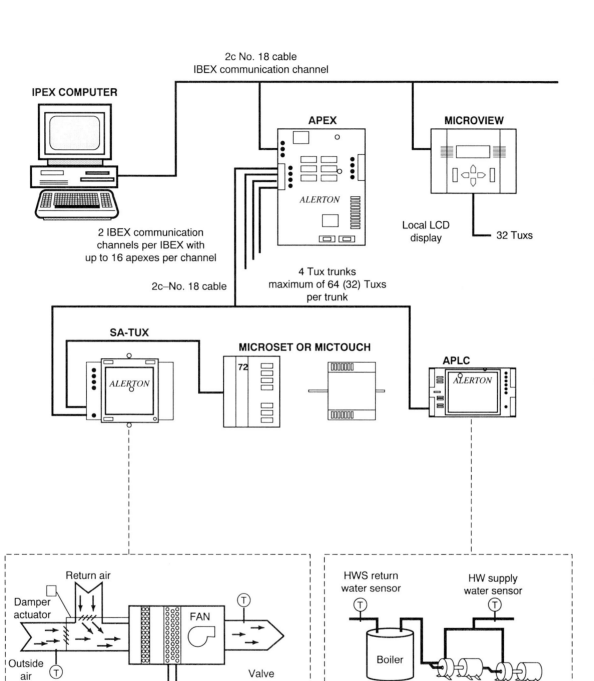

IPEX COMPUTER

2c No. 18 cable
IBEX communication channel

APEX

ALERTON

MICROVIEW

Local LCD
display

32 Tuxs

2 IBEX communication
channels per IBEX with
up to 16 apexes per channel

2c–No. 18 cable

4 Tux trunks
maximum of 64 (32) Tuxs
per trunk

SA-TUX

ALERTON

MICROSET OR MICTOUCH

72

APLC

ALERTON

Return air

Damper
actuator

FAN

T

Outside
air

T

Valve
actuator

HWR

HWS

HWS return
water sensor

T

HW supply
water sensor

T

Boiler

HWS circulating
pumps–lead/lag

Fan Coil/Air Handling Unit

Mechanical System Equipment

Figure 16.7

DDC System Application

capabilities of using input from more than one source, to provide output signals based on combinations of input.

QUESTIONS

16-1. Name the four principal parts of a DDC system as used for controlling an HVAC system.

16-2. Match the term in the column on the left with the description that fits it best in the column on the right by placing the letter found in front of the description in the space provided in front of the term:

A. _____ Sensor   a. processes data from sensors
B. _____ Central processor   b. operating instructions
C. _____ Program   c. operates equipment
D. _____ Actuator   d. monitors conditions

16-3. *True or false:* The sensors used with a DDC system control the mechanical equipment in the system directly.

16-4. *True or false:* The central processing unit in a DDC system both receives signals from the sponsors in the system and originates signals to the actuators in the system.

16-5. The program that actually operates the mechanical equipment in a DDC system is which of the following:
a. Stored in the CPU
b. Part of the operating equipment
c. Stored on disks or tapes
d. Packaged with the computer

16-6. *True or false:* The control signals that operate the actuators in a DDC system come directly from the computer.

16-7. *True or false:* A DDC control system allows programming a system for operation for as long as 1 year.

16-8. What two parts of a DDC control system contain microprocessors for processing input data?

16-9. How many input points can a DDC control system receive data from at one time?
a. 1
b. 3
c. Several
d. 26

16-10. *True or false:* The name *direct digital control* originates from the ability of the system to process digital information but provide analog-type output.

16-11. *True or false:* Only digital sensors are used in a DDC system.

16-12. What type of actuator is used in a DDC control system generally?

APPLICATION EXERCISES

16-1. Complete the schematic diagram for the complete DDC system shown, by drawing in the missing connecting wiring. Write in the names of the control parts used in each section of the system.

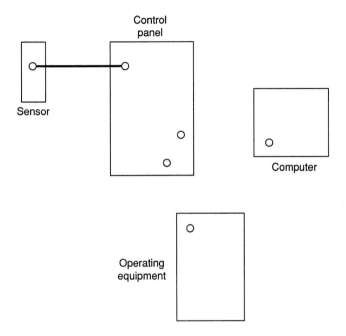

16-2. Complete the two accompanying truth tables to show the proper output for the inputs. Answer the following question:

1. How does Boolean Algebra apply to the way an electronic control system works?

And gate:

Input A	Input B	Output
0	0	
0	1	
1	0	
1	1	

Or gate:

Input A	Input B	Output
0	0	
0	1	
1	0	
1	1	

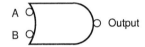

17

Energy Management

INTRODUCTION

Energy management is the control and supervision of the operation of all energy-using equipment in a building for maximum comfort and minimum energy expenditure. The energy we are talking about managing in this book is the energy used for operating mechanical and lighting systems for maintaining a comfortable and safe climate in a building. This is mainly the energy used for heating and cooling the spaces in the building, moving air for comfort, bringing in air for ventilation, and lighting the building. The control system may also include controls related to the safety of the occupants of the building in case of fire or other life-threatening situations.

An energy control system is basically an electronic control system that controls various mechanical components of a comfort system in a building. The major components of the control system are sensors that are used to monitor conditions, a central control panel that receives signals from the sensors and sends signals to operating controls, and the operating controls themselves. A central processing unit (CPU) is part of the central control panel, and other CPUs may be used in other components of the system. A CPU monitors the signals received relative to conditions at various control points in a system, and sends out signals based on the input.

The main energy users in a building are the heating, cooling, ventilation, and lighting systems. The actual percentages of the total energy used by the various components of the system vary depending on the climate in which the building is located and the use of the building with respect to internal heating, cooling, and ventilating loads (Figure 17-1). A building located in a warm climate will normally have a larger percentage of total energy use composed of cooling loads, whereas in a colder climate the larger percentage may be related to heating loads. A building such as a theater auditorium may have a high-density load of people per square foot and thus require a high degree of mechanical cooling or ventilation, whereas a small business office may have a very low-density load of people and thus have little need for cooling or ventilation.

To determine the energy used by the mechanical systems in an existing building, the history of the systems should be reconstructed based on the energy bills received. If the systems are metered separately, this is a simple procedure. If the systems are not metered separately and only figures for the entire building are available, a part of the total energy use shown on the energy bills must be apportioned to each separate energy user by calculating what percentage of the total each represents. In a new building for which no energy use history has been established, the load for each energy-using system must be calculated, and then these sums added to establish the total energy use for the entire building.

Heating Energy

The heating energy needed to heat a building is the energy equivalent of the thermal energy required to offset heat loss from the building when the indoor

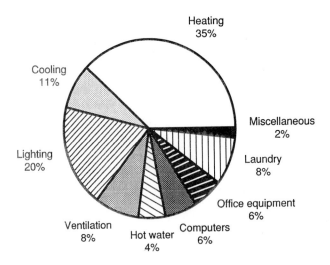

Figure 17-1 The energy used in a building annually can be graphed to show percentage of use by each of the building's energy-using components.

temperature goes below a desired set point (Figure 17-2). Usually, this is caused by the outdoor temperature being less than the desired indoor temperature. The two most common forms of energy used for heating buildings are related to either fuel or electricity. Fuels are burned directly in furnaces or boilers to produce heat, and electricity is used to produce heat through resistance heaters. The amount of fuel or electricity that a heating system uses to heat a building is a function of the building design, use, climate in which the building is located, and heating system design.

Heating load is figured in Btu/h or kWh, depending on which type of heat is used. One Btu (British thermal unit) is approximately the amount of heat required to raise the temperature of 1 pound of water 1 degree Farenheit. The unit Btu/h is the equivalent of 1 Btu per hour.

A watt (W) of electricity is a measurement of electrical energy equivalent to the power produced when 1 volt produces a 1-ampere flow of electricity. A kilowatt (kW) is the equivalent of 1000 watts, and 1 kWh is 1000 watts per hour. In heating loads, kWh is used to indicate the amount of heat that would be produced by 1000 kW of electricity if converted to heat in 1 hour. Finally, 1 kW is equal to 3416 Btu.

Cooling Energy

Cooling energy is the energy expended by an air-conditioning unit or system used to remove excess heat from a building when the inside temperature goes above a desired set point. This usually is a result of the outdoor temperature going above the indoor temperature; other external heat loads, such as solar heat; or by internal heat loads generated by people or equipment.

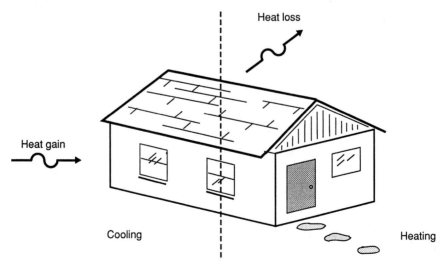

Figure 17-2 The major element in the heat loss or gain of a building is the difference between outdoor and indoor temperature.

Since most cooling equipment uses electricity as a source of energy, cooling energy is generally calculated in kWh. Initial cooling load calculations may be in Btu/h, but for energy-saving determinations, Btu/h is converted to kWh. This is done by dividing Btu/h by 3416, resulting in kWh. For a fuel-burning cooling system both the fuel and electricity will be used as energy, so both must be calculated.

Ventilation Energy

Two energy users are involved in building ventilation. The first is the energy needed to cool or heat the air brought in from outside, to the temperature inside, and the other is the energy used to operate ventilation fans. A small amount of energy is used to operate dampers and damper controls, but this is minimal. In almost all cases, the motors used for ventilation are electrical motors, so the energy used is rated in kWh.

To determine how much energy is used by a ventilation system, the amount of air being moved during a given period of time must be known. This is the air that is brought in from outside the building for ventilation only, usually measured in cubic feet per minute (cfm) of air initially. The cfm value of the air is then converted to pounds, by considering psychrometric properties. The weight of air moved by the ventilation fans is then used to find the amount of work done by the fan motors.

Lights

The energy used to light a building is in the form of electricity. The best way to determine how much energy is being used, or will be used, is to figure the wattage of the lights and the amount of time they will be on during a given period. The wattage and time is used to determine the kWh.

OTHER ENERGY USERS

Other energy users in a building are systems such as food preparation, laundries, elevators, security, and other systems specific to particular types of buildings. If any of these systems provide a significant part of the total energy load of the building, they must be included in the total building load. They can be included separately if they are a significant part of the total building load, or they may be lumped together and included in a miscellaneous category.

ENERGY MANAGEMENT SYSTEMS

An energy management system for a given building is one in which the energy use of components of the various comfort systems in a building is monitored and the operation of the components is controlled to achieve maximum system effect with minimum energy use. The major energy systems include the heating, cooling, ventilation, and lighting systems. Other systems may be included in buildings with specific applications.

Most energy control systems are automated systems using electronic and solid-state technology control components. The major components of a system are sensors that are used to monitor conditions, and a central control panel that receives signals from the sensors and sends signals to the operating controls themselves (Figure 17-3). One central processing unit (CPU) is included in the central control panel, and additional CPUs may be used in other components of the system. Their output is determined by data input and logic (Figure 17-4). The basic units of such control systems are the sensors located at various control points. These are the points from which information concerning system operation is desired. Sensors are available to monitor temperature, humidity, and pressure levels and electrical pulses. Although sensors may be available to monitor other conditions, those noted above are the ones generally monitored in HVAC systems.

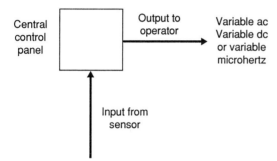

Figure 17-3 The three main components of an energy management control system are the sensors, central control, and the equipment operators.

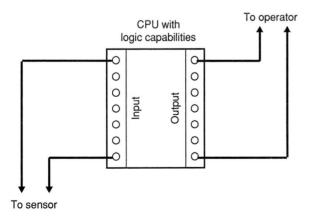

Figure 17-4 Input from the sensors are analyzed logically in the control system central control panel.

Energy Management Systems

Heating System

Heating system controls are designed to turn heating equipment off or on to modulate the output of heat to match the heating load on a building. In an electronic control system the most commonly used sensor that monitors the temperature and provides a modulating signal in relation to the temperature is a device called a thermistor. A thermistor is a solid-state device in which the resistance to electrical flow changes as the temperature changes. A low-voltage dc signal is either sent to the sensor from a central control panel or is originated by the sensor. The signal is modified by the sensor according to the temperature and is then sent to the central control panel. The signal back to the control panel is indicative of the temperature. The modified signal received at the central control panel is analyzed by the CPU in the panel and originates output signals to the system controllers. Heating sensors can be located inside or outside a building, in supply air ducts or hot supply water piping and/or in return air ducts or return water piping.

> **Example** If the outdoor temperature drops suddenly, the control system will reset the indoor temperature set point higher. This causes the heating system to provide additional heat that will be needed in the building. The outdoor sensor responds to the temperature change more rapidly than the indoor sensor will. Another example is use of an outdoor sensor to determine how long a warm-up period is needed in the morning after a building has been on night setback. When the temperature outside is very cold, a longer warm-up time will be needed than when the outside temperature is more temperate.

Cooling System

Cooling system controls are designed to turn cooling equipment off or on to modulate a cooling effect to match the cooling load on a building. In an electronic control system the sensor that monitors the temperature and provides a modulating signal in relation to the temperature is a device called a thermistor. A thermistor is a solid-state device in which the resistance to electrical flow changes as the temperature changes. In energy-saving control systems, cooling control is used in several ways.

> **Example** One of the special applications in which both heating and cooling controls are used in energy systems is when they are used to prevent both heating and cooling systems operating at the same time, in opposition to each other, when multiple heating/cooling units are used to control the temperature in large open spaces. This application is often used in large retail or wholesale houses and in large manufacturing applications. Another application is for sensing the temperature at different control points within a refrigeration system to control unloading of the system to match more closely the cooling load on a building.

Unloading an air-conditioning unit involves an arrangement of refrigeration piping that allows some of the hot refrigerant gas leaving the compressor to be returned to the suction line ahead of the compressor instead of going through the rest of the system. This effectively reduces the cooling capacity of the unit.

In many buildings heat pumps are used for heating and cooling. The heat pump itself is an energy-saving device in the heating cycle, but by using water-to-air heat pumps and controlling the flow of water from one part of a building to another, the entire system can be made to operate more efficiently when both heating and cooling.

Example An example of an energy-saving control system using heat pumps is in a large office building with both some northern and some southern exposure (Figure 17-5). During certain times of the year, areas with the northern exposure may require heating while areas of southern exposure need cooling. With proper controls, the heat pumps on the north side of the building can take heat out of the water that goes to them, thus heating building spaces. This cools the water that is returned to the hydronic system. This cooler water is then circulated to the south side of the building, where it is distributed to heat pumps there. The heat pumps in this part of the building pick up heat from the air and return it to the water. The cycle can then start over. The control system senses the need for cooling or heating in the various zones in the building, and controls water pumps

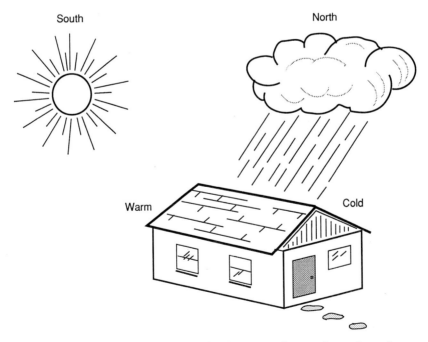

Figure 17-5 A large building will often have control zones that each require different treatment to provide comfort in the entire building.

and valves to direct the flow of water between the two general zones. The temperature in each building space is controlled by thermostats in the spaces.

Ventilation System

Ventilation air is the air brought into a building from outside to freshen the air being circulated. The amount of air is determined from the number of occupants of a building, and to some extent by their activities. Ventilation system control is usually either by dampers (Figure 17-6) or by control of ventilation fan motors. Ventilation air uses energy directly through the use of electricity to run the ventilation fan motors, but it is also a source of energy use as a result of the need to heat or cool the air that is brought in from outside. This energy is secondary to that for the fan motors, but it is a significant cause of total ventilation air energy use. Proper application of controls can help to reduce energy use from both causes.

Example In many building ventilation systems, one fan is used for both general air circulation and for ventilation air circulation. When this is the case, ventilation air control is by means of dampers. In ventilation systems using ventilation fans with their own motors, the motors may either be cycled on and off or the speed of the ventilation fan modulated for control of ventilation air. In many applications, ventilation air to a building is shut off when a building is not occupied, but turned on when it is. This provides a direct energy savings.

Electrical energy can usually be saved for both ventilation and general air handling if motor speed control is used in place of damper operation to provide variable airflow. Control sensors are used to monitor either airflow or air pressure in building spaces to determine how much air is entering the spaces. The sensors provide a low-voltage dc signal to the control system control center. The signal is proportional to the airflow or air pressure. In the control center the control signal actuates output signals that control the speed controllers on the fan motors involves. The fan motors turn faster or slower to provide just the amount of air out of the blowers to satisfy system demand.

A special case of ventilation air control is that in which outdoor air is used to help cool a building when the outdoor air temperature is below the cooling set

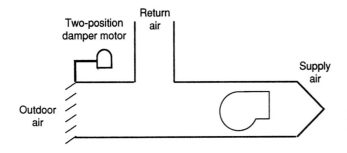

Figure 17-6 Ventilation air is often controlled by a damper on the outdoor air inlet.

point. The control package to accomplish this is called an economizer system (Figure 17-7).

Lighting System

Controlling the lights in a building to economize on energy use is generally a matter of turning lights on only when they are needed and of controlling the light intensity to the level needed at a given time. Traditionally, the design of lighting systems for buildings in which people are studying or working has called for enough lights to produce about 100 candlepower at the working level by the lighting system alone, and for the lights to be turned on during all hours when the building may be occupied. But 100 candlepower is a higher level of lighting than is needed for many activities, many building spaces are not occupied at all times, and most buildings have some light entering from outside at some times. If the lighting system and the lighting control system are properly designed to conserve lighting energy, the system will provide in each building space the amount of light required for specific activities, the lights will be on only during occupied periods, and light from windows and/or special lighting panels will be used to help light the spaces.

To help achieve the requirements noted above, light-sensitive sensors are installed in interior building spaces, and through the lighting control system,

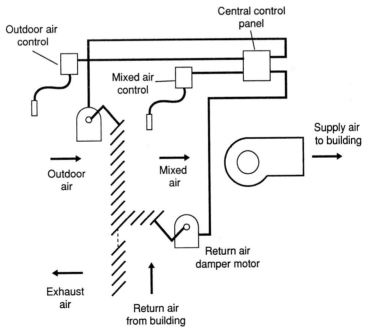

Figure 17-7 An economizer control system requires control of return air and outdoor dampers, and in some cases, of exhaust dampers also.

Energy Management Systems

regulate the lights in the building spaces to provide a constant level of light intensity. If light coming in from outside increases, some lights are turned off; if the light coming in from outside decreases, additional lights are turned on.

In an energy measurement system photo-sensitive sensors are usually used for sensing light intensity (Figure 17-8). The photosensor in a lighting system receives a low-voltage dc signal from a central control panel and modifies the signal according to the light to which it is exposed. The modified signal actuates relays to turn lights off and on to provide just the level of light intensity desired in each building space.

PROGRAM

The basis of an energy-saving control system is the computer program that is used to monitor and control the equipment in a building system. The program is written by a trained programmer, on a computer, in a language that another computer can "understand." The program is saved on tapes, diskettes, or disks. The program identifies parameters within which the various mechanical components of a building comfort system should operate.

Example If the temperature of a space in a building drops to 4°F below the set point on the thermostat, the heating unit for the space turns on. If the temperature in the space rises to 2°F below the set point, the heating unit is shut off. If the temperature at the thermostat goes to 4°F above the set point, the cooling unit is turned on. If the temperature drops to 2°F above the set point, the cooling unit is turned off. There is always a noll zone between heating and cooling during which neither heating nor cooling will be on (Figure 17-9).

To control the ventilation air in a building the program would have param-

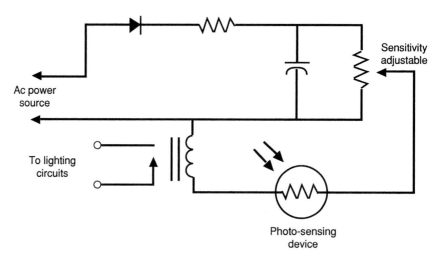

Figure 17-8 The basic control of a lighting system depends on a photo-sensing device that measures light intensity.

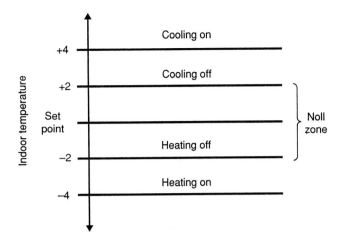

Indoor temperature

+4 — Cooling on

+2 — Cooling off

Set point

Heating off

−2

Heating on

−4

Noll zone

Figure 17-9 Temperature above and below the set point at a control point in a building space determines the operation of heating and cooling for the space.

eters written into it to maintain outdoor air dampers in a minimum position to allow for ventilation for people during all occupied periods. The same dampers would be closed during nonoccupied periods. The occupied and nonoccupied periods are usually identified by time. The ventilation fans will be turned on to run continuously during occupied periods but will be turned off during nonoccupied periods. The ventilation fan will normally be operative whenever heating or cooling is called for, except during setback periods. Setback periods are those when the set-point temperature at the thermostat is set back to a lower temperature for heating and a higher one for cooling. This is usually done during unoccupied periods.

The energy management control program also has parameters written into it to control the light level in various spaces in a building. The light level, measured in candlepower, is controlled either by controlling the voltage to lights or by turning banks of light off and on as needed.

One of the most energy-saving features of a control system is the control of electrical equipment to make sure that the total electrical energy used at any one time does not exceed a maximum level set by the local electrical company. Such a maximum limit is called a demand limit. If the electrical load on a building exceeds this limit, the customer is penalized through a higher rate per kWh for the electricity used. Designing the control system to avoid exceeding the limit automatically saves energy.

To make sure that the electrical load on a building does not exceed the demand limit, sensors on the electrical distribution lines to various pieces of mechanical equipment in the building monitor the amount of electrical energy being used by each piece. That information is conveyed to the control system central control by output signals from the sensors. In the central control panel the signals are analyzed through the logic circuits in the CPU, and output signals from the central control are used to control the mechanical equipment. When the total electrical demand approaches the demand limit, the central control shuts down some of the mechanical equipment. The equipment that is shut down will be that which can most easily be spared at the time.

Program

SUMMARY

An energy management control system in a building constitutes a set of controls that control the operation of HVAC equipment in a building to provide a given level of comfort with the most economical energy expenditure. The primary energy users in a building are the heating, cooling, ventilation, and lighting systems. Controlling the use of energy for heating a building is primarily a function of providing just that amount of heat required to offset the heat loss of the building at any given time. Similarly, the saving of energy for cooling is the provision of just the cooling effect required at any given time.

Ventilation of a building is the introduction of fresh outside air to the air circulating in a building. This is to dilute the air in the building to keep it fresh and clean. The introduction of outside air to a building uses energy in two ways: the energy used to run ventilation fans, and the energy used to heat or cool the outdoor air so that its temperature will be the same as the indoor air temperature. The control system used to control the amount of air brought in for ventilation includes various sensors that indicate occupancy or nonoccupancy of a building.

To save energy for the lighting system in a building, sensors are used that indicate such conditions as occupancy and light levels. Lights are turned off when a building space is not occupied, and are controlled to maintain a level of light intensity in the space that is just right for use of the space.

Control of an energy control system is by means of a computer program that is stored on disks or tape. The program provides parameters of operation for the various pieces of mechanical equipment in a building.

REVIEW QUESTIONS

17-1. What are the two main objectives to be achieved when using an energy management control system?

17-2. Name the four primary energy-using systems found in most buildings.

17-3. Name at least three energy-using systems that are found in many buildings in addition to those you listed in question 2.

17-4. Name three primary components of an electronic energy control system.

17-5. What causes a heating system to be cycled on and off? (Answer in relation to the thermostat set point.)

17-6. What is the main reason that the temperature in a building deviates from the set point?

17-7. Name two additional reasons that the temperature in a building goes above the set point besides those listed in question 6.

17-8. Name two other causes of temperature drop in a building besides those listed in question 6.

17-9. What is the name of the electronic device used as a sensor for heating and cooling systems, in which electrical resistance is proportional to temperature?

17-10. Name two different ways in which the introduction of ventilation air in a building contributes to energy use.

17-11. What is the name of the damper and control system used to help cool a building when the outdoor temperature is cooler than the set point?

17-12. What two primary goals are to be achieved when designing a lighting control system?

17-13. Choose which of the four answers below best fits the question: Where does an energy control system computer program come from?

    (a) It comes with the computer.

    (b) It is written by a computer programmer.

    (c) You buy it at EggHead Computer.

    (d) It just happens naturally.

APPLICATION EXERCISES

17-1. Using history or estimated values, draw a pie chart showing the percentage use of energy in a small commercial building in your area as assigned by your instructor. Fill in the data as called for on the graph.

    1. Building name:

    2. Building address:

    3. Source of data:

17-2. A small commercial building has annual energy use as follows:

Gas fuel: 864,000 Btu/h

Electricity: 316.2 kWh

    (a) What is the total annual load in Btu (that is, Btu/yr)?

    (b) What is the total annual load in kWh (that is, kWh/yr)?

17-3. Make a survey of the lighting system and of all electrical energy users in a typical commercial building in your area. From the wattage found, estimate the annual use of lights and appliances. Show the energy use for each category (lights, and appliances) for 1 year.

    Building name:

    Building address:

    Source of information:

    Annual energy use for lights _____

    Annual energy use for appliances _____

17-4. On the accompanying diagram, write in the names of the controls and operators in the spaces provided, and show in single lines the control wiring for a typical ventilation damper system.

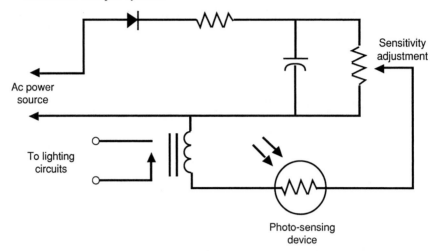

17-5. On the accompanying diagram, write in the names of the controls and operators in the spaces provided, and draw in the control wiring for a typical economizer damper package.

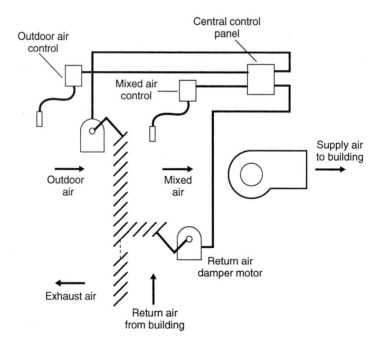

Control Glossary

There are many terms peculiar to the fields of control and control systems as there are in any field of science or technology. These terms are often used in reference to the application of controls and control systems. It is necessary for a control technologist to be familiar with the terms related to those systems.

A

Action: generally refers to the action of a controller and defines what is done to regulate the final control element to effect control. Types of action include on–off, proportional, rate, and reset.

Actuator: control element that translates the controller output into an action by the final control element.

Ambient: related to conditions surrounding a process.

Analog: continuous range of values, such as temperature, pressure, and so on. Contrast with **digital.**

Automatic control system: any combination of automatic controllers connected in closed loops with one or more processes.

Automation: employment of devices that automatically control one or more functions in a process.

B

Bellows: pressure-sensing element consisting of a convoluted metal cylinder closed at one end. Pressure difference between the outside and inside of the bellows causes it to expand or contract along its axis.

Bimetallic element: temperature-sensitive element of a bimetallic control. The bimetal is composed of two or more metal alloys mechanically bonded so that they bend one way when heated and the other way when cooled.

Bridge circuit: electrical circuit network in which the value of an unknown component is obtained by balancing one circuit against another. It consists of a detecting device and four resistances connected in series/parallel circuits to form a diamond.

C

Calibration: procedure laid down for determining, correcting, or checking the absolute values corresponding to the graduations on a measuring instrument.

Celsius scale: centigrade temperature scale, for which the freezing point of water is 0° and the boiling point is 100°.

Central processing unit: that part of a computer which contains the arithmetic and logic functions that process programmed instructions.

Closed loop: arrangement of components to allow system feedback. If the measured value differs from the desired value, a corrective signal is sent ot the final control element to bring the controlled variable to the proper value.

Coefficient of linear expansion: increase in unit length per degree of temperature rise.

Computer: data processor that can perform substantial computation, including numerous arithmetic or logic operations, without human intervention.

Computer program: series of instructions or statements in a form acceptable to a computer prepared to achieve a specific result (software).

Control agent: process energy or material that is manipulated to hold a control medium at its desired value. In heating a building with hot water, water is the control agent.

Control point: (1) value of a controlled variable that under any fixed set of conditions, an automatic controller operates to maintain; (2) physical point at which a controller is located in a system.

Control pressure (pneumatic controllers): output pressure as modified by a controller.

Control range: range between the upper and lower output limits of a controller.

Control system: assemblage of control apparatus coordinated to execute a planned set of control functions.

Controlled device: instrument that receives the controller's output signal and regulates the flow of the control agent. It is functionally divided into two parts: *Actuator:* receives the output signal and converts it into force. *Regulator:* valve body or damper that regulates the flow of the control agent.

Controlled medium: substance such as air in a room that is temperature, humidity, or pressure controlled, and which is affected by the control agent and by heat gains or losses to surrounding areas.

Controlled variable: quantity or condition of the controlled system that is directly measured and controlled.

Controller: instrument that measures the controlled variable and responds by producing an output signal that is proportional to the difference between the set point and the control point.

D

Data processing: execution of a systematic sequence of operations performed on data. Synonymous with information processing.

Derivative: rate at which something changes, or quantities increase or decrease. Also the name of a control action in which the output is proportional to the input's rate of change.

Deviation: difference between the instantaneous value of the controlled variable and the set point.

Differential: applies to two-position (on–off) controller action. It is the smallest range of values through which the controlled variable must pass in order to move the controller output from its "on" to its "off" position (or vice versa).

Digital: refers to data expressed in numerical format.

Direct acting: output signal changing in the same direction the controlled or measured variable changes (e.g., an increase in the controlled or measured variable results in an increased output signal).

E

Economizer control: in air-handling systems, the control system that selects outside air, recirculated air, or a mixture of the two for the most energy-efficient cooling.

Electromechanical control system: control system that uses relatively low voltage electricity for control signals, and electric and mechanical devices as operators and controllers.

Electronic control system: control system that uses low-voltage dc electricity as a signal medium, with electronic operators and controllers.

Enthalpy control: control device that senses the total heat content of air. It works by sensing both humidity and temperature.

F

Farenheit scale: temperature scale on which the freezing point of water is 32° and the boiling point is 212°.

Feedback: part of a closed-loop system that provides information about a given condition for comparison with the desired condition.

Feedback signal: signal that is returned to the input of the system and compared with the reference signal to obtain an actuating signal that returns the controlled variable to the desired value.

H

Heat anticipation: variation of two-position action in which the "on" periods are prematurely shortened. A heating element in a room thermostat wired into the control circuit so that it is on when the thermostat calls for heat and off when the thermostat calls for cooling.

Hydraulic control system: control system that uses fluid pressure as a signal medium and also as an energy transfer system. It usually uses mechanical control devices for fluid management.

Hygrometer: instrument for measuring moisture content.

I

Inclined tube manometer: manometer with one leg at an angle, permitting the scale on that arm to be expanded for more precise readings of low pressure.

Input: incoming signal to a control unit or system.

Instrument: used broadly to connote a device incorporating measuring, indicating, recording, controlling, and/or operating abilities.

Integral mode: control mode that produces control action that is proportional to the accumulation of error over a period of time.

Interface: shared boundary. An interface might be a hardware component to link two devices or it might be a portion of storage or registers accessed by two or more computer programs.

L

Lag: refers to delay, and is expressed in seconds or minutes. Caused by conditions such as capacitance, inertia, resistance, and dead time, either separately or in combination.

Load: demand on the operating resources of a system. *Heating:* rate of heat loss of the space being controlled. *Cooling:* rate of heat gain to the space being controlled.

Logic: in microprocessors, logic uses symbols to represent quantities and relationships identifiable as gates or switching circuits; these can be arranged to perform logical functions necessary for equipment operation.

M

Manometer: gauge for measuring pressure of gases and vapors.

Manual controller: controller having all its basic functions performed by devices that are operated by hand.

Medium: solid or fluid through which a force or effect is conveyed.

Microprocessor: central processing unit (CPU) of a microcomputer; this section generally contains the arithmetic logic unit and the control logic unit.

N

Normally closed: applies to a controlled device that opens when a signal is applied to it.

Normally open: applies to a controlled device that closes when a signal is applied to it.

O

Offset: sustained deviation of the controlled variable from the set point. (This characteristic is inherent in proportional controllers that do not incorporate reset action.) Offset is caused by load changes.

On—off control action: occurs when a final control element is moved from one or two fixed positions to the other with a very small change of controlled variable. Same as *two-position action*.

Open loop: system in which no comparison is made between the actual value and the desired value of a process variable.

Operating pressure range: stated high- and low-pressure values of pneumatic pressure required to produce full-range operation when applied ot a pneumatic intelligence-transmission system, a pneumatic motor operator, or a pneumatic positioning relay.

Output: outgoing signal of a transmitter or control unit.

P

Photelectric cell: device whose electrical properties undergo a change when the device is exposed to light.

Pilot: auxiliary mechanism that actuates or regulates another mechanism.

Pneumatic control system: control system that uses low-pressure air as a signal medium, with mechanical control devices used to control the air and as operators.

Potentiometer: measures by comparing the difference between known and unknown electrical potentials. To measure process control variables by means of a potentiometer, these variables, such as temperature, pressure, flow, and liquid level, must first be translated into electrical signals that vary proportionally with changes in the variable.

Pressure-sensing element: part of a pressure transducer that converts the measured pressure into a mechanical motion.

Printed circuit board: insulating board with circuits printed on one side and electronic control components mounted on the other side. The electronic components are connected to the circuits through the board. The components are arranged to perform logic or control functions as part of a control system.

Printout: recording by a typewriter or printer.

Programmable control system: control system in which set points can be entered in relation to various control functions for specific time, temperature, humidity, pressure, or other variables desired.

Proportional action: produces an output signal proportional to the magnitude of the input signal. In a control system proportional action produces a value correction proportional to the deviation of the controlled variable from the set point.

Psychrometer: instrument for measuring humidity.

Psychrometric chart: chart showing graphically the relationship between all properties of dry air and the moisture contained in it. Properties such as temperature, humidity, and dew point are featured.

R

Range: difference between the maximum and minimum values of physical output over which an instrument is designed to operate normally.

Readout: visual or printed display of measured values.

Relative humidity: ratio between the amount of water vapor actually present in the air and the maximum amount of water vapor the air could hold at the same temperature.

Relay: device that enables the energy in one circuit to be controlled by the energy in another.

Reset action: control action that produces a corrective signal proportional to the length of time the controlled variable has been away from the set point. Also called *integral action.*

Reverse acting: output signal changing in the opposite direction of the controlled or measured variable changes (e.g., an increase in the controlled or measured variable results in a decreased output signal).

S

Semiconductor: material, such as silicon or germanium, that has a greater resistance to current flow than a conductor, but not as great a resistance as an insulator.

Sensing element: part of a controller that is in contact with the medium being measured and which responds to changes in the medium.

Sensitivity: (1) ratio of change of output to chnage of input; (2) least signal input capable of causing an output signal having desired characteristics.

Set point: position at which the control-point setting mechanism is set. This is the same as the desired value of the controlled variable.

Signal: information conveyed from one point in a transmission or control system to another. Signal changes usually call for action or movement.

Software: set of computer programs, procedures, and associated documentation concerned with the operation of a computer (e.g., compilers, library routines, manuals, and flowcharts).

Solenoid: electromagnet having an energizing coil approximately cylindrical in form, and an armature whose motion is reciprocal within and along the axis of the coil.

Specific heat: ratio of the thermal capacity of any substance to the thermal capacity of water.

Specific humidity: amount of water vapor per unit mass of moist air.

T

Temperature: relative hotness or coldness of a body as determined by its ability to transfer heat to its surroundings. A measurement of degree of heat; an indication of the relative motion of the molecules in a material.

Thermistor: resistor whose resistance varies with temperature in a definite desired manner. Used in circuits to compensate for temperature variation, to measure temperature, or as a nonlinear circuit element.

Thermocouple: pair of dissimilar conductors so joined that an electromotive force is developed by thermoelectric effects when the two junctions are at different temperatures.

Thermometer: device for measuring temperature.

Throttling range (proportional controllers): change in controlled variable required to move the controlled device(s) from one extreme limit of travel to the other.

Transducer: instrument that converts an input signal of one energy form into an output signal of another energy form (e.g., electrical input to pneumatic output).

Transistor: semiconductor amplifying device.

U

U-tube manometer: form of manometer used for pressure measurement.

V

Variable: process condition, such as pressure, temperature, flow, or level, which is susceptible to change and which can be measured, altered, and controlled.

Velocity meter: instrument for measuring fluid velocity or rate of flow.

W

Wet bulb: temperature sensor covered with a wet cloth. It is used with a dry-bulb thermometer for humidity measurement.

Wheatstone bridge: four-arm bridge, all arms of which are predominately resistive.

Working range: desired controlled or measured variable values over which a system operates.

Index

295

Air-conditioning unit controls (*cont.*)
  types of, 108–9
  ventilation units, 122–25
Airflow control:
  electromechanical, 50–51
    airflow proving switches, 51
    sail switch, 51
  electronic, 54–55
  pneumatic, 52–53
Airflow sensing, 75–77
  fans, 75
  filters, 75
  ventilation, 75–77
Air movement, 15–19
  filter material, 17–18
  stratification, 15, 17
Ambient temperature, 35
Automated controls, 249, 250

B

Bellows actuators, 195–96
Bimetallic element, 48
Bimetallic sensors, 69–70
Blade switches, 87, 130
Blower, 16
  controls, 232–33, 238–40
Boolean algebra, 263
Bridge circuit, 39–41, 72–73, 151–52
Btu, definition of, 26

C

Central processing unit (CPU),
    microprocessor, 250, 251
Chiller, 119
Circuit breakers, 87, 88, 130–31
Cold contacts, 99–100
Combustion safety controls, 7, 86, 96–
    97
  electronic control systems, 235–36
  gas-fired heating units, 96–97
  oil-fired heating units, 99–100
Combustion safety reset, 137–38
Comfort, 12–25
  control points, 21–23
  heating/cooling media, 19–21
  indoor climate, 12–19
Contactors, 77–78, 110, 111, 154, 177
Control devices:
  electric controls, 152–56
  electronic control, 221–25

Control function, 68–84
  actuators, 79–81
  feedback, 81
  sensing conditions, 68–77
  signals, 77–78
Controllers, 47–55, 68, 152–53
  electromechanical, 48–51
  electronic, 53–55, 222
  pneumatic, 51–53
Control offset, 237
Control operation, 147–52
  digital control, 147
  proportional controls, 150–52
  stepped controls, 147–50
Control points, 21–23
  definition of, 21
  locations of, 21–23
    combining, 23
  outside, 21
  and sensing of solar radiation, 22–
    23
Control systems, 7
  electrical, 7, 81, 129–39
  electronic, 7, 217–25
  operation of, 34–44
    digital control, 34–36
    proportional control, 39–44
    stepped control, 36–39
Control theory, 26–45
  control system operation, 34–44
  equipment operation, 26–34
Cooling energy, 273–74
Cooling system controls, 276–78
Cooling thermostats, 110–11
Couple, 71, 96
Cylinder bypass, 175–76
Cylinder unloading, 176

D

Damper actuators, 61
Damper control, 178–79
Damper control systems, electronic, 241–
    43
Dampers, 75
DDC, *See* Direct digital control (DDC)
Defrost controls, air-conditioning units,
    114–15
Dehumidification controls, 122
Detente device, 166
Diaphragm actuators, 60

Ventilation system controls, 278–79

Ventilation units, 122–25
  dampers, 123
  mixed-air control, 123–24
  outdoor temperature control, 124–25
  outside-air control, 123

return-air control, 123

Voltage regulators, 135

W

Warm-air heating systems, 3

Water-curtain humidification system, 121

Wet-bulb depression, 75

Wet-bulb temperature, 13